Dictionary Learning in Visual Computing

Synthesis Lectures on Image, Video, & Multimedia Processing

Editor
Alan C. Bovik, *University of Texas, Austin*

The Lectures on Image, Video and Multimedia Processing are intended to provide a unique and groundbreaking forum for the world's experts in the field to express their knowledge in unique and effective ways. It is our intention that the Series will contain Lectures of basic, intermediate, and advanced material depending on the topical matter and the authors' level of discourse. It is also intended that these Lectures depart from the usual dry textbook format and instead give the author the opportunity to speak more directly to the reader, and to unfold the subject matter from a more personal point of view. The success of this candid approach to technical writing will rest on our selection of exceptionally distinguished authors, who have been chosen for their noteworthy leadership in developing new ideas in image, video, and multimedia processing research, development, and education.

In terms of the subject matter for the series, there are few limitations that we will impose other than the Lectures be related to aspects of the imaging sciences that are relevant to furthering our understanding of the processes by which images, videos, and multimedia signals are formed, processed for various tasks, and perceived by human viewers. These categories are naturally quite broad, for two reasons: First, measuring, processing, and understanding perceptual signals involves broad categories of scientific inquiry, including optics, surface physics, visual psychophysics and neurophysiology, information theory, computer graphics, display and printing technology, artificial intelligence, neural networks, harmonic analysis, and so on. Secondly, the domain of application of these methods is limited only by the number of branches of science, engineering, and industry that utilize audio, visual, and other perceptual signals to convey information. We anticipate that the Lectures in this series will dramatically influence future thought on these subjects as the Twenty-First Century unfolds.

MPEG-4 Beyond Conventional Video Coding: Object Coding, Resilience, and Scalability
Mihaela van der Schaar, Deepak S Turaga, and Thomas Stockhammer
2006

Modern Image Quality Assessment
Zhou Wang and Alan C. Bovik
2006

Biomedical Image Analysis: Tracking
Scott T. Acton and Nilanjan Ray
2006

Recognition of Humans and Their Activities Using Video
Rama Chellappa, Amit K. Roy-Chowdhury, and S. Kevin Zhou
2005

Dictionary Learning in Visual Computing

Qiang Zhang and Baoxin Li

ISBN: 978-3-031-01125-2   paperback
ISBN: 978-3-031-02253-1   ebook

DOI 10.1007/978-3-031-02253-1

A Publication in the Springer series
SYNTHESIS LECTURES ON IMAGE, VIDEO, & MULTIMEDIA PROCESSING

Lecture #18
Series Editor: Alan C. Bovik, *University of Texas, Austin*
Series ISSN
Print 1559-8136   Electronic 1559-8144

Dictionary Learning in Visual Computing

Qiang Zhang
Samsung

Baoxin Li
Arizona State University

SYNTHESIS LECTURES ON IMAGE, VIDEO, & MULTIMEDIA PROCESSING #18

ABSTRACT

The last few years have witnessed fast development on dictionary learning approaches for a set of visual computing tasks, largely due to their utilization in developing new techniques based on sparse representation. Compared with conventional techniques employing manually defined dictionaries, such as Fourier Transform and Wavelet Transform, dictionary learning aims at obtaining a dictionary adaptively from the data so as to support optimal sparse representation of the data. In contrast to conventional clustering algorithms like K-means, where a data point is associated with only one cluster center, in a dictionary-based representation, a data point can be associated with a small set of dictionary atoms. Thus, dictionary learning provides a more flexible representation of data and may have the potential to capture more relevant features from the original feature space of the data. One of the early algorithms for dictionary learning is K-SVD. In recent years, many variations/extensions of K-SVD and other new algorithms have been proposed, with some aiming at adding discriminative capability to the dictionary, and some attempting to model the relationship of multiple dictionaries. One prominent application of dictionary learning is in the general field of visual computing, where long-standing challenges have seen promising new solutions based on sparse representation with learned dictionaries. With a timely review of recent advances of dictionary learning in visual computing, covering the most recent literature with an emphasis on papers after 2008, this book provides a systematic presentation of the general methodologies, specific algorithms, and examples of applications for those who wish to have a quick start on this subject.

KEYWORDS

Dictionary Learning, Sparse Coding, Sparse Representation, Compressive Sensing, Matrix Completion, Image Compression, Image Denoising, Image Inpainting, Image Demosaicing, Image Super-resolution, Image Segmentation, Background Subtraction, Blind Source Separation, Saliency Detection, Visual Tracking, Face Recognition

To my parents – QZ

To the memory of my father,
and to my mother, wife and daughters – BL

Contents

Acknowledgments

It is a pleasure to acknowledge many colleagues who have made this time-consuming book project possible and enjoyable. In particular, many current and past members of the Visual Representation and Processing Group at Arizona State University have worked on various aspects of sparse learning and its applications in visual computing. Their efforts have supplied ingredients for insightful discussion related to the writing of the book, and thus are greatly appreciated.

We thank Professor Rama Chellappa of University of Maryland for providing constructive comments and especially for pointing us to a few early efforts demonstrating the deep roots of many recent techniques. We are also grateful to Professor Huan Liu at ASU, who never hesitated to offer his advice and share his valuable experiences whenever the authors needed them. The insightful comments of Dr. Nasser Nasrabadi from ARL have strenghtened our discussion around sparsity models, which are sincerely appreciated.

This work stemmed from efforts in several projects sponsored by NSF and ARO, whose supports are greatly appreciated (the views/conclusions in this book are solely of the authors and do not necessarily reflect those of the sponsors).

We are grateful to Morgan & Claypool and particularly Executive Editor Diane D. Cerra for her help and patience throughout the writing of this book.

Last, but foremost, we thank our families for their unwavering support during this fun project, for their understanding and tolerance of many weekends and long nights spent on the book by the authors. We dedicate this book to them, with love.

Qiang Zhang and Baoxin Li
March, 2015

Figure Credits

Figure 3.2   From: Bristow, H., Eriksson, A., and Lucey, S. (2013). Fast convolutional sparse coding. In *Computer Vision and Pattern Recognition (CVPR), 2013 IEEE Conference on*, pages 391-398. Copyright © 2013 IEEE. Used with permission.

Figure 3.3   From: Jiang, Z., Lin, Z., and Davis, L. S. (2011). Learning a discriminative dictionary for sparse coding via label consistent k-svd. In *Computer Vision and Pattern Recognition (CVPR), 2011 IEEE Conference on*, pages 1697-1704. Copyright ©2011 IEEE. Used with permission.

Figure 3.5   From: Hawe, S., Seibert, M., and Kleinsteuber, M. (2013). Separable dictionary learning. In *Computer Vision and Pattern Recognition (CVPR), 2013 IEEE Conference on*, pages 438-445. Copyright © 2013 IEEE. Used with permission.

Figure 3.6   From: Jenatton, R., Mairal, J., Bach, F. R., and Obozinski, G. R. (2010). Proximal methods for sparse hierarchical dictionary learning. In *Proceedings of the 27th International Conference on Machine Learning (ICML-10)*, pages 487-494. Used with permission.

Figure 3.7   Based on: Zhou, M., Chen, H., Ren, L., Sapiro, G., Carin, L., and Paisley, J. W. (2009). Nonparametric bayesian dictionary learning for sparse image representations. In *Advances in Neural Information Processing Systems*, pages 2295-2303.

Figure 3.8   From: Ding, X., He, L., and Carin, L. (2011). Bayesian robust principal component analysis. *Image Processing, IEEE Transactions on*, 20(12):3419-3430. Copyright © 2011 IEEE. Used with permission.

Figure 3.9   Based on: He, L., Qi, H., and Zaretzki, R. (2013). Beta process joint dictionary learning for coupled feature spaces with application to single image super-resolution. In *Computer Vision and Pattern Recognition (CVPR), 2013 IEEE Conference on*, pages 345-352. IEEE.

Figure 4.1	From: Aharon, M., Elad, M., and Bruckstein, A. (2006). K-SVD: An algorithm for designing overcomplete dictionaries for sparse representation. *IEEE Transactions on Signal Processing*, 54 (11), 4311-4322. Copyright ©2006 IEEE. Used with permission.
Figure 4.2	From: Bryt, O. and Elad, M. (2008). Compression of facial images using the k-svd algorithm. *Journal of Visual Communication and Image Representation*, 19(4):270-282. Copyright © 2008. Elsevier. Used with permission.
Figure 4.4	From: Dong, W., Li, X., Zhang, D., and Shi, G. (2011a). Sparsity-based image denoising via dictionary learning and structural clustering. In *Computer Vision and Pattern Recognition (CVPR), 2011 IEEE Conference on*, pages 457-464. Copyright ©2011 IEEE. Used with permission.
Figure 4.5	Mairal, J., Bach, F., Ponce, J., and Sapiro, G. (2009a). Online dictionary learning for sparse coding. In *Proceedings of the 26th Annual International Conference on Machine Learning*, pages 689-696. Used with permission.
Figure 4.7	From: Mairal, J., Bach, F., Ponce, J., Sapiro, G., and Zisserman, A. (2009b). Non-local sparse models for image restoration. In *Computer Vision, 2009 IEEE 12th International Conference on*, pages 2272-2279. Copyright ©2009 IEEE. Used with permission.
Figure 4.8	From: Yang, S., Liu, Z., Wang, M., Sun, F., and Jiao, L. (2011b). Multitask dictionary learning and sparse representation based single-image super-resolution reconstruction. *Neurocomputing*, 74(17):3193-3203. Copyright ©2011 Elsevier. Used with permission.
Figure 4.9	From: Zhang, Q., Zhou, J., Wang, Y., Ye, J., and Li, B. (2014). Image cosegmentation via multitask learning. In *British Machine Vision Conference*. Used with permission.
Figure 4.10	From: Wright, J., Ganesh, A., Rao, S., Peng, Y., and Ma, Y. (2009a). Robust principal component analysis: Exact recovery of corrupted low-rank matrices by convex optimization. In *Proc. of Neural Information Processing Systems*, volume 3. Used with permission.
Figure 4.11	From: Yan, J., Zhu, M., Liu, H., and Liu, Y. (2010). Visual saliency detection via sparsity pursuit. *Signal Processing Letters, IEEE*, 17(8):739-742. Copyright © IEEE. Used with permission.

Figure 4.12   From: Mei, X. and Ling, H. (2009). Robust visual tracking using ℓ 1 minimization. In *Computer Vision, 2009 IEEE 12th International Conference on*, pages 1436-1443. Copyright © 2009 IEEE. Used with permission.

CHAPTER 1

Introduction

The last few years have witnessed fast development on dictionary learning approaches for many visual computing tasks, largely due to the utilization of such approaches in developing new techniques based on sparse representation. Compared with conventional approaches relying on manually defined dictionaries, such as Fourier Transform, Wavelet Transform, and so on, dictionary learning approaches typically utilize over-complete matrices to support (sparse) representation of the input space and thus the learning task aims at deriving a set of vectors (dictionary atoms) adaptively from the input data under some optimality criterion. Also, in contrast to conventional, clustering-based representations such as K-means, where a data point is associated with only one cluster, dictionary-based sparse representation allows a point to be associated with a small set of dictionary atoms. By adapting the atoms to the input, dictionary learning can potentially provide a more compact meanwhile more flexible representation of the input data, hence capturing more relevant information from the original feature space for further analysis tasks. Among the most well-known algorithms for dictionary learning is the K-SVD method [Aharon et al., 2005], which has inspired many variations or extensions, including discriminative dictionary learning and learning with hierarchical dictionaries.

Dictionary learning has been applied in various visual computing tasks, and great success has been demonstrated in applications like image denoising, image super-resolution, face recognition, image classification, and so on. With an emphasis on the most recent literature, this book aims at summarizing recent advances of dictionary learning in visual computing tasks, including presentation of the fundamental ideas, discussion of algorithms and their features, and application examples. In doing so, the book intends to provide a quick reference for a reader interested in studying and applying dictionary-learning-based techniques for visual computing tasks.

Despite the plethora of research efforts on dictionary learning in the recent literature, the basic idea of dictionary-based representation is by no means new. In fact, many recent techniques around dictionary learning may be traced back to relevant or similar ideas in much earlier literature. For example, the orthogonal matching pursuit algorithm popularized by Tropp and Gilbert [2007] may have its root in Pati et al. [1993]. Also, while recent years have seen many new discriminative dictionary learning algorithms (e.g., Zhang and Li [2010a]), an early effort already attempted separability-based basis selection in the wavelet domain. The remainder of this chapter will review two classes of conventional approaches to signal/data representation, i.e., transforms based on orthonormal bases and clustering. These approaches, while not necessarily being called "dictionary-based" in the original literature, enjoy an interpretation with well-defined dictionaries. Moreover, due to their wide application and good "literacy," understanding these approaches from a dictionary-based perspective serves to facilitate the study of more recent techniques as well as their deep root in the earlier literature.

A note on notations: In this book, we will use uppercase letters in boldface for matrices (e.g., $\mathbf{X}$), lowercase letters for vectors (e.g., $\mathbf{x}$), and uppercase letters for constants (e.g., X). The i_{th} column of matrix $\mathbf{X}$ will be written as $\mathbf{x}_i$, and $\mathbf{x}^j$ is for the j_{th} row, where i or j can be either a scalar, or a set for selecting multiple columns or rows. We also use $\{\cdots\}$ to represent a set and $[\cdots]$ to build a matrix from a set of column vectors.

1.1 ORTHOGONAL DICTIONARIES IN TRANSFORMS

In discrete Fourier transform (DFT), a signal $\mathbf{x} \in \mathbb{R}^{N \times 1}$ can be represented by its Fourier coefficient $\mathbf{y} \in \mathbb{R}^{N \times 1}$ as

$$\mathbf{x}_j = \frac{1}{\sqrt{N}} \sum_{k=1}^{N} [y_k e^{\frac{-i2\pi(N-k+1)(j-1)}{N}}], \ \forall j = 1, \cdots, N \tag{1.1}$$

which is the inverse Fourier transform. If we build a matrix $\mathbf{F} \in \mathbb{R}^{N \times N}$ as $\mathbf{F}_j^k = \frac{1}{\sqrt{N}} e^{\frac{-i2\pi(k-1)(j-1)}{N}}$, we have

$$\mathbf{x} = \sum_{k} [\mathbf{F}_k \mathbf{y}_k] = \mathbf{F}\mathbf{y}. \tag{1.2}$$

That is, we may say that the signal $\mathbf{x}$ is represented by $\mathbf{y}$ under the dictionary $\mathbf{F}$. To compute the representation of the given signal under $\mathbf{F}$, we can use

$$\mathbf{y}_k = \frac{1}{\sqrt{N}} \sum_{j=1}^{N} [y_k e^{\frac{-i2\pi(k-1)(j-1)}{N}}], \ \forall k = 1, \cdots, N. \tag{1.3}$$

As the dictionary $\mathbf{F}$ is orthonormal (i.e., $\mathbf{F}^H \mathbf{F} = \mathbf{I}$, where H is the conjugate transpose operator), we can also compute it as $\mathbf{y} = \mathbf{F}^H \mathbf{x}$.

With this dictionary-based view of DFT, one may think about using different dictionaries in place of the above Fourier bases to obtain other types of representations for the given signal/data. Indeed, different dictionaries lead to different transforms. We describe a few such examples below.

The Sinusoid Family. This includes the DFT introduced above, Discrete Cosine Transform (DCT), and Discrete Sine Transform (DST), all of which use sinusoid functions to form the bases. The dictionaries for DFT, DCT, and DST, respectively, can be computed as

$$\mathbf{F}_j^k = \frac{1}{\sqrt{N}} e^{\frac{-i2\pi(k-1)(j-1)}{N}} \tag{1.4}$$

$$\mathbf{C}_j^k = \frac{1}{\sqrt{N}} \cos\left[\frac{\pi}{N}(j - \frac{1}{2})(k-1)\right] \tag{1.5}$$

$$\mathbf{S}_j^k = \frac{1}{\sqrt{N}} \sin\left[\frac{\pi}{N+1} jk\right]. \tag{1.6}$$

As an example, Fig. 1.1 illustrates the DCT dictionaries of sizes 8, 16, and 256, respectively. The computational complexity of the DFT, DCT, and DST is $O(N \log N)$ for an input of $\mathbb{R}^{N \times 1}$.

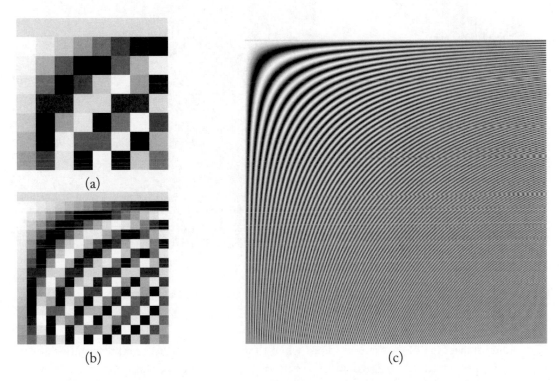

(a)

(b)   (c)

Figure 1.1: Some examples of dictionaries for DCT of size 8 (a), 16 (b), and 256 (c).

The Walsh-Hadamard Transform (WHT). In this case, the bases can be computed iteratively as:

$$\mathbf{B}_1 = \begin{bmatrix} +1 & +1 \\ +1 & -1 \end{bmatrix}; \quad \mathbf{B}_n = \begin{bmatrix} +\mathbf{B}_{n-1} & +\mathbf{B}_{n-1} \\ +\mathbf{B}_{n-1} & -\mathbf{B}_{n-1} \end{bmatrix}, \tag{1.7}$$

where $\mathbf{B}_n \in \mathbb{R}^{2^n \times 2^n}$ is the basis matrix of order n. Figure 1.2 illustrates the WHT dictionaries of size 8, 16, and 256, respectively. The computational complexity is also $O(N \log N)$ for $N = 2^n$. One advantage of the WHT over the sinusoid family is that its computation only involves integer operation, and thus it can be made to run much faster. In addition, it has lower memory footprint.

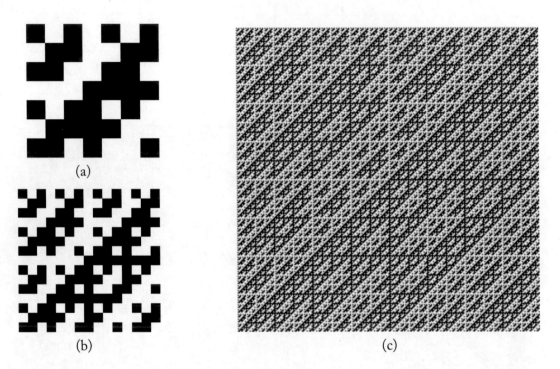

(a)

(b)

(c)

Figure 1.2: Some examples of dictionaries for WHT of size 8 (a), 16 (b), and 256 (c).

Dictionaries are of particular interest when they can support *sparse* representation (or more generally, approximation) of signals/data. Intuitively, a sparse representation is one in which only a small number of coefficients (under a given dictionary) is sufficient to capture the given data accurately. It has been found that many natural signals acquired from sensory systems can be sparse-represented under a dictionary from the above exemplar transforms. The representation of a signal under these dictionaries can be computed very efficiently, with the complexity being $O(N \log N)$ for an input of $\mathbb{R}^{N \times 1}$. It is evident that these dictionaries are static: they are pre-defined, independent of any given data. A natural question would be: Will it be beneficial to design a dictionary adaptive to the data? Not surprisingly, the answer to this question alludes to the general problem of dictionary learning, which is the subject of this book.

Algorithm 1 The K-means algorithm

Input : input signals $\mathbf{X}$
Output : cluster centers $\mathbf{D}$ and cluster assignment α

1: Initialize cluster centers $\mathbf{D}$.
2: **while not** converged **do**
3:      Compute the Euclidean distance of each data point $\mathbf{x}_i$ to each cluster center $\mathbf{d}_j$
4:      For each data point, update the assignment α_i by assigning it to the closest center
5:      Update each cluster center by computing the mean of the data assigned to this cluster center
6:      Check stop criterion
7: **end while**

1.2   DICTIONARIES IN CLUSTERING ALGORITHMS

There is an intriguing relation between sparse representation and clustering (or vector quantization), which has previously been mentioned in the literature [Cheney and Cline, 2004, Engan et al., 2000, Kreutz-Delgado et al., 2003]. In clustering, a set of descriptive vectors $\{d_k\}_{k=1}^{K}$ is learned, and each sample is represented by one of those vectors (the one closest to it, usually measured in the ℓ^2 distance). Take the K-means algorithm [MacQueen et al., 1967] as an example, which aims to find a set of cluster centers $\{\mathbf{d}_j \in \mathbb{R}^{d \times 1} | j = 1, \cdots, K\}$ and clustering assignments $\{\alpha_i \in \mathbb{R}^{K \times 1} | i = 1, \cdots, N\}$ from a set of data points $\{\mathbf{x}_i \in \mathbb{R}^{d \times 1} | i = 1, \cdots, N\}$. The algorithm can be formulated as the following optimization problem, where the notion of "dictionary" has become explicit:

$$\mathbf{D}, \alpha : \min_{\mathbf{D}, \alpha} \frac{1}{2} \|\mathbf{X} - \mathbf{D}\alpha\|_F^2 \text{ s.t. } \|\alpha_i\|_1 = 1, \|\alpha_i\|_0 = 1, 0 \leq \alpha_i \leq 1, \forall i = 1, \cdots, N, \qquad (1.8)$$

where $\mathbf{D} = [\mathbf{c}_1, \cdots, \mathbf{c}_K]$ (with K being the number of clusters), and $\mathbf{X}$ and α are similarly defined. The constraints $\|\alpha_i\|_1 = 1, \|\alpha_i\|_0 = 1, 0 \leq \alpha_i \leq 1$ ensure that there is only one element of α_i taking value 1 and all others being 0.

We may think of this as an extreme case of sparse representation, where only one atom of the dictionary is allowed in representing the signal and the corresponding coefficient must be one. A typical way of solving Eq. 1.8 is given in Algorithm 1.

In Algorithm 1, we assume that the Euclidean distance is used to determine the association of a data point and a cluster center. If we use the ℓ_1 distance, i.e., considering instead $\min_{\mathbf{D}, \alpha} \frac{1}{2} \|\mathbf{X} - \mathbf{D}\alpha\|_1$, we need to update the cluster center with the median of the data assigned to this cluster center. For the stop criterion, three types of conditions are typically used: (1) the algorithm has reached a pre-set maximal number of iterations; or (2) the cluster centers become (nearly) static. Regarding the initialization of $\mathbf{D}$, there are several possibilities, including assigning $\mathbf{D}$ randomly, setting it to some randomly-chosen training data points, or running an initial clustering on a randomly selected subset of the training data.

Although being widely used in many fields, the K-means algorithm has several inherent limitation. First, the algorithm can only guarantee convergency to a local optimum and the final result heavily depends on the initialization. To this end, different ideas have been explored, such as running several trials with different initialization and picking the best clustering. Second, the number of clusters, i.e., K, needs to be manually specified, which is often not an easy task when little is known about the distribution of the data *a priori*. One possibility is to try different values for K, and then pick the K which leads to most dramatic reduction/increase to the the average distance of the data points to their cluster centers. Last but not the least, each data point is allowed to be represented by only one cluster center, which is very restrictive. In practice, depending on the actual data distribution, it may be beneficial to consider schemes that allow a data point to be associated with multiple centers.

Besides the basic K-means technique, some other types of clustering algorithm have also been proposed, such as hierarchical clustering, including agglomerative clustering and divisive clustering [Szekely and Rizzo, 2005, Zhang et al., 2012d, 2013d], density-based clustering and its variants [Ester et al., 1996][Ankerst et al., 1999], and more sophisticated ways of doing statistical clustering (e.g., Expectation-Maximization clustering [Yang, 2005], and the Dirichlet mixture process [Rasmussen, 1999]). These algorithms still suffer from one or more of the limitations mentioned above for the basic K-means technique.

Instead of requiring each data to be represented by only one cluster center as in K-means, in sparse representations as discussed in this book, each sample is represented as a linear combination of several vectors $\{d_k\}_{k=1}^{K}$. Thus, sparse representations can be referred to as a generalization of the clustering problem. [Coates and Ng, 2011] presented a comparative study of sparse coding and vector quantization (clustering). Experiments on CIFAR, NORB, and Caltech 101 datasets were used to show that sparse coding outperforms vector quantization in image classification tasks.

Since the K-means algorithm (sometimes called the generalized Lloyd algorithm-GLA [Gersho and Gray, 1992]) is the most commonly used procedure for training in the vector quantization setting, it is natural to consider generalizations of this algorithm when turning to the problem of dictionary training, leading to the basic idea of the K-SVD algorithm, which will be presented in the following chapter.

The remaining of this book is organized as follows. In Chapter 2, we introduce several common ways of formulating the sparse learning problem. These formulations will be frequently used in subsequent chapters when we discuss various dictionary learning algorithms and their applications. In Chapter 3, we introduce some of the most well-known dictionary learning algorithms from the recent literature, by organizing them into five categories, namely, reconstructive dictionary learning, discriminative dictionary learning, joint learning of multiple dictionaries, online dictionary learning, and statistical dictionary learning. Applications of these learning algorithms in solving various visual computing tasks are presented in Chapter 4, covering signal compression, signal recovery, image super-resolution, segmentation, classification, saliency detection, and visual tracking. Finally, in Chapter 5, we conclude the book with an instructive example of applying dictionary learning to the face recognition problem.

CHAPTER 2

Fundamental Computing Tasks in Sparse Representation

In both the case of conventional orthonormal dictionaries and the case of over-complete dictionaries, often sought is sparsity in representation under certain properly defined dictionaries. Hence, sparse learning and dictionary learning will be often interchangeably used in this book. In general, sparsity of representation is introduced to seek a trade-off between goodness-of-fit (e.g., reconstruction errors) and simplicity (i.e., sparsity in this context) of the representation. Depending on the actual formulation of the problem, the meaning of sparsity can often appear to be different, and the corresponding solution will also vary. In this chapter, we discuss several common ways of formulating the sparse learning problem, along with basic ideas behind the solutions for these formulations. The details of various algorithms for solving the learning problem are to be elaborated in the next chapter.

2.1 DICTIONARY-BASED SPARSE REPRESENTATION

Sparse learning has seen many applications in visual computing, e.g., compressed sensing [Donoho, 2006], image denoising [Elad and Aharon, 2006], image compression [Elad and Aharon, 2006], image super resolution [Yang et al., 2012], face recognition [Wright et al., 2009c], visual tracking [Liu et al., 2011a], image classification [Yang et al., 2009], visual saliency [Yan et al., 2010], action recognition [Guha and Ward, 2012], and so on. In these applications, a basic formulation for sparse learning can often be obtained by incorporating a sparsity term into the original learning formulation, resulting in the following problem:

$$\alpha : \min_{\alpha} f(\alpha) + g(\alpha) \text{ s.t. } \mathbf{x} \in \Omega, \tag{2.1}$$

where α is a vector, $f(\cdot)$ is the term from traditional learning methods, $g(\cdot)$ is the sparsity term and Ω is the feasible set. Different terms have been proposed for $g(\cdot)$, e.g., ℓ_0, which measures the number of nonzero elements [Chen et al., 1991], i.e., $\|\alpha\|_0 = \sum_i (\alpha_i \neq 0)$. However, ℓ_0 is not convex and typically makes it NP-hard to find a solution. Instead, the ℓ_1 norm, which is the sum of absolute values of the elements, or $\|\alpha\|_1 = \sum_i |\alpha_i|$, is often used.

One of the simplest and most common examples is:

$$\alpha : \min_{\alpha} \frac{1}{2} \|\mathbf{x} - \mathbf{D}\alpha\|_2^2 \text{ s.t. } \|\alpha\|_0 \leq \tau. \tag{2.2}$$

Algorithm 2 The matching pursuit algorithm

Input : input signal $\mathbf{x}$, dictionary $\mathbf{D}$
Output : sparse coefficient α

1: **while not** converged **do**
2:    Compute the correlation of $\mathbf{x}$ to each dictionary atom $\mathbf{d}_i$ as $\frac{\mathbf{x}^T \mathbf{d}_i}{\|\mathbf{x}\|_2 \|\mathbf{d}_i\|_2}$
3:    Pick the atom $\mathbf{d}_i$ which has largest correlation and set $\alpha_i = \frac{\mathbf{x}^T \mathbf{d}_i}{\|\mathbf{x}\|_2 \|\mathbf{d}_i\|_2}$
4:    Compute the signal residual by subtracting the selected atom $\mathbf{x} = \mathbf{x} - \alpha_i \mathbf{d}_i$
5:    Check stop criterion
6: **end while**

This problem (nonconvex) can be efficiently solved via matching pursuit (MP) [Mallat and Zhang, 1993]: In each iteration, a column of $\mathbf{D}$, $\mathbf{d}_i$, is selected, which has the largest correlation with $\mathbf{x}$, and then we make the update $\mathbf{x} = \mathbf{x} - \frac{\mathbf{x}^T \mathbf{d}_i}{\|\mathbf{d}_i\|_2^2}$. This procedure is repeated until τ columns are selected or $\|\mathbf{x}\|_2$ is smaller than some value. The algorithm is also described in Algorithm 2.

One drawback of the above matching pursuit algorithm is that, as the dictionary atoms are not all orthogonal to each other, a dictionary atom could be selected more than once, which reduces the performance of the algorithm. To alleviate this issue, orthogonal matching pursuit [Pati et al., 1993, Tropp and Gilbert, 2007] (OMP) was developed, where the major difference is that, after a new dictionary atom is selected, the coefficients for all the selected atoms are updated, by computing the orthogonal projection of the signal onto the selected atoms, which is described in Algorithm 3. OMP generally provides a more sparse coefficient than MP. A detailed study of convergence behaviors of different types of greedy algorithm for sparse coding was presented in [Barron et al., 2008]

If we replace ℓ_0 by ℓ_1 and apply the Lagrange multiplier technique, we will obtain the following convex problem:

$$\alpha \ : \ \min_{\alpha} \frac{1}{2} \|\mathbf{x} - \mathbf{D}\alpha\|_2^2 + \gamma \|\alpha\|_1 \tag{2.3}$$

$$\alpha \ : \ \min_{\alpha} \frac{1}{2} \|\mathbf{x} - \mathbf{D}\alpha\|_2^2 \ \text{s.t.} \ \|\alpha\|_1 \leq \tau \tag{2.4}$$

$$\alpha \ : \ \min_{\alpha} \|\alpha\|_1 \ \text{s.t.} \ \|\mathbf{x} - \mathbf{D}\alpha\|_2^2 \leq \epsilon. \tag{2.5}$$

The problem in Eq. 2.3 is also known as least absolute shrinkage and selection operator (or LASSO).

It has been proven in Candes et al. [2006] that, if $\mathbf{D}$ satisfies uniform uncertainty principle (with unit-norm column) and if the to-be-recovered vector α_0 is sufficiently sparse, then the

Algorithm 3 The orthogonal matching pursuit algorithm

Input : input signal $\mathbf{x}$, dictionary $\mathbf{D}$
Output : sparse coefficient α

1: Let Ω be the index set of the selected dictionary atoms and initialize as $\Omega = \emptyset$
2: **while not** converged **do**
3:      Compute the correlation of $\mathbf{x}$ to each dictionary atom $\mathbf{d}_i$, which is not in Ω, as $\frac{\mathbf{x}^T \mathbf{d}_i}{\|\mathbf{x}\|_2 \|\mathbf{d}_i\|_2}$
4:      Pick the atom $\mathbf{d}_i$ which has largest correlation and include in into the set $\Omega = \Omega \cup \{i\}$
5:      Update the coefficient $\alpha_\Omega = (\mathbf{D}_\Omega^T \mathbf{D}_\Omega)^{-1} (\mathbf{D}_\Omega^T \mathbf{x})$
6:      Compute the signal residual by subtracting the selected atom $\mathbf{x} = \mathbf{x} - \mathbf{D}_\Omega \alpha_\Omega$
7:      Check stop criterion
8: **end while**

solution to Eq. 2.5 is close to $\mathbf{x}_0$:

$$\|\alpha - \alpha_0\|_2 \leq C\epsilon \tag{2.6}$$

where C is some constant (related to the property of $\mathbf{D}$) and ϵ the tolerant reconstruction error in Eq. 2.5.

Many types of dictionaries satisfy uniform uncertainty principle (UUP [Needell and Vershynin, 2009]), e.g., random matrices with entries independently randomly sampled, matrices with columns randomly selected from Fourier basis, and even matrices with columns randomly selected from orthogonal matrices.

Many tools have been proposed for solving the sparse learning problems presented above. Following is an incomplete list: GPSR [Chen et al., 1991], ℓ_1 magic [Candes and Romberg, 2005], NESTA [Becker et al., 2011a], $\ell_1$ls [Koh et al., 2007], SparseLab [Donoho et al., 2005], SpaRSA [Wright et al., 2009d], ℓ_1 Homotopy [Asif and Romberg, 2013], FISTA [Sastry and Ma, 2010], TFOCS [Becker et al., 2011b], approximate message passing [Kamilov et al., 2012], Bregman iterative regularization [Yin et al., 2008], YALL1 [Yang and Zhang, 2011], and so on. A benchmark comparison for some of these algorithms was reported in Sastry and Ma [2010].

Though being useful and popular, ℓ_0 or ℓ_1 based sparsity priors have several limitations, e.g., they fail to capture complex sparse structures. As a result elastic net has been proposed [Zou and Hastie, 2005], which can be written as $\lambda_2 \|\alpha\|_2^2 + \lambda_1 \|\alpha\|_1$, where the ℓ_1 part ($\|\alpha\|_1$) generates a sparse model, and the ℓ_2 part ($\|\alpha\|_2^2$) encourages group selection. The elastic net is also convex, which can also be solved efficient by many solvers, e.g., LARS-EN [Zou et al., 2006].

In many application problems, similar signals may come in group. Sparse coding on those signals separately may be not only insufficient but also ineffective, as the similarity of the signal is not exploited. Considering this, group sparse coding was proposed in Bengio et al. [2009], where similar signals are considered as a group and the signals of the same group are encouraged to be sparsely represented by a similar set of the dictionary atoms. To do this, a group sparsity inducing norm $\|\alpha\|_{p,q} = (\sum_i (\sum_j |\alpha_i^j|^p)^{\frac{q}{p}})^{\frac{1}{q}}$ was proposed to replace the ℓ_0 norm or ℓ_1 norm.

$$\alpha : \min_\alpha \frac{1}{2} \sum_i \|\mathbf{x}_i - \mathbf{D}\alpha_i\|_2^2 + \gamma \|\alpha\|_{p,q}, \tag{2.7}$$

where α_i^j is the j_{th} sparse coefficient for i_{th} sample and $\|\alpha\|_{p,q}$ is usually referred as $\ell_{p,q}$ mixture norm or ℓ_q/ℓ_p norm. To induce the group sparsity constraint, $(2, 1)$ and $(\infty, 0)$ are the typical choices for (p, q).

In the $\ell_{p,q}$ norm, all the signals are assumed to belong to the same group, which may not be true for some scenarios. Other sparse priors have been proposed to capture more complex structures in the signal. Let $\{g_i : i = 1, ...G\}$ be the groups in α, with $g_i \in \{1, 2, 3, ..., K\}$ and α_{g_i} being the subvectors of α with indices from g_i, then the group sparsity of the vector α can be written as [Yuan and Lin, 2006]

$$\|\alpha\|_G = \sum_i \|\alpha_{g_i}\|_2 \tag{2.8}$$

Here $\|\alpha\|_G$ enforces the sparsity on the selected groups, rather than elements within the groups. In Friedman et al. [2010], a $\|\alpha\|_1$ term was further added to $\|\alpha\|_G$, to enforce sparsity also on the input vector, which is named sparse group lasso.

The group structure may form hierarchies (such as in wavelet coefficients of a signal). Accordingly, hierarchical sparsity prior is also proposed (e.g., Baraniuk et al. [2010], Zhao et al. [2009]). The hierarchical structure may be represented in trees, where if an element is selected, its parent element must also be selected (i.e., the zero-tree structure). An example of dictionary learned with hierarchical sparsity prior is shown in Fig. 3.6.

The sparse structure can also be generally represented by a graph, with both the hierarchical structure and group structure being special cases [Huang et al., 2011]. In graph sparsity, each node of the graph represents an index or a subset of indices in the vector, and the edges between two nodes indicate the corresponding indices belong to a group. This structure can often be found in images, where pixels form the nodes and the edges represent neighbor relations among the pixels.

2.2   SPARSE REPRESENTATION WITH MATRICES

Sparse learning method can also be generalized to matrix. Addition to the number of nonzero elements or sum of absolute values of the elements, the rank of a matrix has been given a lot of interest when it comes to spare learning in matrices. A good example is illustrated by the matrix completion problem [Candes and Plan, 2009], where the missing entries of a matrix are to be recovered under the assumption that, the rank of the matrix $\mathbf{X} \in \mathbb{R}^{m \times n}$ is low:

$$\mathbf{Y} : \min_{\mathbf{Y}} \text{rank}(\mathbf{Y}) \text{ s.t. } \mathbf{Y}_i^j = \mathbf{X}_i^j \, \forall (i, j) \in \Omega, \tag{2.9}$$

where rank($\cdot$) is the rank of the matrix and Ω is the set of observed entries (accordingly $\bar{\Omega}$ is the set of missing entries). $\mathbf{Y} \in \mathbb{R}^{m \times n}$ is the recovered matrix, which is required to be identical to the input matrix $\mathbf{X}$ for the observed entries, namely exact matrix completion. In some cases, we may not require exact matrix completion but the error $P_\Omega(\mathbf{X} - \mathbf{Y})$ to be small, with the sampling operation $P_\Omega(\cdot) : \mathbb{R}^{m \times n} \to \mathbb{R}^{m \times n}$ defined as:

$$[P_\Omega(\mathbf{X})]_i^j = \begin{cases} \mathbf{X}_i^j & \text{if } (i, j) \in \Omega \\ 0 & \text{otherwise} \end{cases} \tag{2.10}$$

Accordingly, we can reformulate the problem as:

$$\mathbf{Y} \; : \; \min_{\mathbf{Y}} \text{rank}(\mathbf{Y}) \text{ s.t. } \|P_\Omega(\mathbf{Y} - \mathbf{X})\|_F^2 \leq \epsilon \tag{2.11}$$

$$\mathbf{Y} \; : \; \min_{\mathbf{Y}} \text{rank}(\mathbf{Y}) + \frac{\mu}{2} \|P_\Omega(\mathbf{Y} - \mathbf{X})\|_F^2. \tag{2.12}$$

The rank of a matrix can be measured as the number of linear independent columns or rows of the matrix, whichever is smaller. It can also be measured as the number of nonzero singular values for the matrix [Candes and Plan, 2009], i.e., $\text{rank}(\mathbf{X}) = \|\Sigma\|_0$, where $\mathbf{X} \to \mathbf{U}\Sigma\mathbf{V}^T$ is the singular value decomposition. However, the rank measurement is not convex, thus its convex relaxation, trace norm (or nuclear norm), is often used [Cai et al., 2008]. The nuclear norm or trace norm of a matrix is defined as the sum of the singular value of the matrix, i.e., $\|\mathbf{X}\|_* = \|\Sigma\|_1$ (please note that for Σ, all of its elements are non-negative and its off-diagonal elements are zero.). Accordingly, we have:

$$\mathbf{Y} \; : \; \min_{\mathbf{Y}} \|\mathbf{Y}\|_* \text{ s.t. } \mathbf{Y}_i^j = \mathbf{X}_i^j \, \forall (i, j) \in \Omega \tag{2.13}$$

$$\mathbf{Y} \; : \; \min_{\mathbf{Y}} \|\mathbf{Y}\|_* \text{ s.t. } \|P_\Omega(\mathbf{Y} - \mathbf{X})\|_F^2 \leq \epsilon \tag{2.14}$$

$$\mathbf{Y} \; : \; \min_{\mathbf{Y}} \|\mathbf{Y}\|_* + \frac{\mu}{2} \|P_\Omega(\mathbf{Y} - \mathbf{X})\|_F^2. \tag{2.15}$$

It has been proven that [Candes and Plan, 2009], if the rank of $\mathbf{X}$ is 1 then under mild conditions

$$k \geq C\mu^4 \max(m, n) \log^2 \max(m, n), \tag{2.16}$$

where $k = \|\Omega\|$ is the number of observed entries in the input matrix $\mathbf{X}$, C, and μ are some positive constants. For a matrix with a rank larger than 1, we have

$$k \geq C\mu^2 r \max(m, n) \log^6 \max(m, n), \tag{2.17}$$

where r is the rank of $\mathbf{X}$, C and μ are some positive constant, then with high probability that the problems in Eq. 2.9 and Eq. 2.13 have the identical solution, or,

$$\|\hat{\mathbf{Y}} - \mathbf{Y}\|_F \leq 4\sqrt{\frac{(2 + p)\min(m, n)}{p}}\delta + 2\delta, \tag{2.18}$$

where $\hat{\mathbf{Y}}, \mathbf{Y}$ are solutions for Eq. 2.9 with $p = \frac{k}{m \times n}$ and δ is some small constant.

This problem has been studied for addressing famous Netflix challenge [Meka et al., 2009], foreground segmentation in video [Wright et al., 2009a], face recognition [Zhang and Li, 2012], and so on.

The problem in Eq. 2.13 can also be solved in its dual form:

$$\mathbf{W} : \max_{\mathbf{W}} \langle \mathbf{W}, \mathbf{X} \rangle \text{ s.t. } \|\mathbf{W}\|_2 \leq 1, \tag{2.19}$$

where $\mathbf{W}$ is the Lagrange multiplier. The constraint $\|\mathbf{W}\|_2 \leq 1$ is semi-positive definite, which can be written as:

$$\begin{bmatrix} \mathbf{I}_m & \mathbf{W} \\ \mathbf{W}^T & \mathbf{I}_n \end{bmatrix} \succeq 0 \tag{2.20}$$

with $\mathbf{I}_m \in \mathbb{R}^{m \times m}$ being the identity matrix. Thus the problem in Eq. 2.19 can be solved via semi-definite programming (SDP) . However, the high complexity of SDP implies that Eq. 2.19 can only be solved at small scale. In practice, a much more efficient algorithm called singular value thresholding (SVT) [Cai et al., 2010] is used. To solve Eq. 2.19, SVT iteratively uses a shrinkage operator defined below:

$$\text{shrink}_\tau(\mathbf{X}) = \mathbf{U} \mathbf{S}_\tau(\Sigma) \mathbf{V}^T, \tag{2.21}$$

where $\mathbf{X} \to \mathbf{U} \Sigma \mathbf{V}^T$ is singular value decomposition and $\mathbf{S}_\tau(\cdot)$ is the following soft thresholding operator:

$$\mathbf{S}_\tau(x) = \begin{cases} x - \tau & \text{if } x \geq \tau \\ 0 & \text{if } -\tau \leq x \\ \tau - x & \text{otherwise.} \end{cases} \tag{2.22}$$

The singular value thresholding algorithm is presented in Algorithm 4. For the stop criterion, two types of conditions have been proposed: the maximal number of iterations reaches, and the normalized residual $\frac{\|\mathbf{P}_{\Omega}(\mathbf{Y}-\mathbf{X})\|_F}{\|\mathbf{P}_{\Omega}(\mathbf{X})\|_F}$ is smaller than the predefined value. The parameter ρ controls the convergence speed, which is typically selected from $(1, 2]$; τ is the step size, which is initialized as $\delta = 1.2 \frac{m \times n}{k}$ for input matrix $\mathbf{X} \in \mathbb{R}^{m \times n}$ [Cai et al., 2010].

2.3   SPARSE REPRESENTATION VIA STATISTICAL LEARNING

Sparse learning can be interpreted under the Bayesian framework [Ji et al., 2008]. Given the noisy observation $\mathbf{y} \in \mathbb{R}^{d \times 1}$, the task to recover the sparse coefficient $\mathbf{x} \in \mathbb{R}^{k \times 1}$ under some projection matrix (or dictionary) $\mathbf{D} \in \mathbb{R}^{d \times k}$ can be formulated as following problems:

$$\alpha \quad : \quad \max_{\alpha} p(\alpha | \mathbf{x}, \mathbf{D}) \tag{2.23}$$
$$\max_{\alpha} p(\mathbf{x} | \mathbf{D}, \alpha) p(\alpha).$$

Algorithm 4 The singular value thresholding algorithm

Input : locations of samples Ω, sampled entries $P_\Omega(X)$, parameters τ, δ, ρ
Output : recovered Y

1: Initialize $Z = \text{ceil}(\frac{\tau}{\delta\|P_\Omega(X)\|_2})\delta P_\Omega(X)$
2: **while not** converged **do**
3:    $Y = \text{shrink}(Z)$
4:    $Z = Z + \delta P_\Omega(X - Y)$
5:    $\delta = \delta\rho$
6:    Check stop criterion
7: **end while**

If we assume that the noises in the noisy observation x follows a zero-mean Gaussian distribution with variance σ^2, i.e.,

$$p(x|D, \alpha) = p(x - D\alpha) = (2\pi\sigma^2)^{-\frac{k}{2}} e^{-\frac{\|x - D\alpha\|_2^2}{2\sigma^2}} \tag{2.24}$$

and that the coefficient α obeys the Laplacian distribution, i.e.,

$$p(\alpha) = (\frac{\lambda}{2})^d e^{-\lambda \sum_i |\alpha_i|} \tag{2.25}$$

then we have

$$\begin{aligned}
\alpha \quad : \quad &\max_\alpha (2\pi\sigma^2)^{-\frac{k}{2}} e^{-\frac{\|x - D\alpha\|_2^2}{2\sigma^2}} (\frac{\lambda}{2})^d e^{-\lambda \sum_i |\alpha_i|} \\
&\max_\alpha (2\pi\sigma^2)^{-\frac{k}{2}} (\frac{\lambda}{2})^d e^{-\frac{\|x - D\alpha\|_2^2 - 2\sigma^2\lambda \sum_i |\alpha_i|}{2\sigma^2}} \\
&\min_\alpha \frac{1}{2}\|x - D\alpha\|_2^2 + \gamma\|\alpha\|_1,
\end{aligned} \tag{2.26}$$

where to derive Eq. 2.26, we can simply take the negative logarithm.

By comparing Eq. 2.26 and Eq. 2.3, we observe that these two equations are identical, by setting $\gamma = \lambda\sigma^2$. This also indicates how to set the parameter γ: if the observation is expected to be very noisy (or σ^2 is large), then γ should be set to some large value to allow large reconstruction error $\frac{1}{2}\|x - D\alpha\|_2^2$ in obtaining the sparse coefficient α; or if the coefficient x is known to be very sparse (or λ is large), then γ should also be set to some large value to obtain the sparse coefficient.

The spike-and-slab sparse coding (S3C) method was proposed in Goodfellow et al. [2012] for large scale feature learning. In S3C, the observation x is generated according to the following process[1] (which is also illustrated in Fig. 2.1):

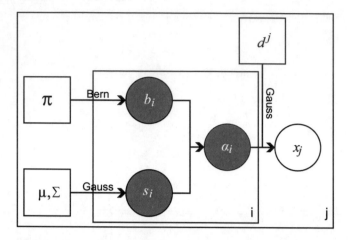

Figure 2.1: A graph showing how the visible variable is drawn. The circle represents the observable variable, the circle in gray for the hidden variable and the rectangle for the parameters.

$$
\begin{aligned}
\mathbf{x}_j &\sim N(\mathbf{x}|\mathbf{D}^j\alpha, \sigma_j^2) \\
\alpha_i &= \mathbf{b}_i\mathbf{s}_i \\
\mathbf{b}_i &\sim \text{Bernoulli}(\pi_i) \\
\mathbf{s}_i &\sim N(0, \Sigma),
\end{aligned}
$$

where π is the parameter controlling the sparseness of the hidden variable (sparse coefficient) α, (μ, Σ) controls the value of the nonzero elements of α, and σ controls the noise in the observation α. $\mathbf{s}$ (the slab) or the Gaussian distribution decides the magnitude of the element in α, while the binary vector $\mathbf{b}$ (the spike) or the Bernoulli distribution controls the locations of nonzero elements of α.

Compared with the sparse coding methods presented in Sec. 2.1, S3C has several potential benefits. In ℓ_1 norm based sparse coding methods (e.g., Eq. 2.3), the sparse coefficient ($\mathbf{x}$) is not merely encouraged to be sparse; it is also encouraged to be close to zero, even if it is active (nonzero). This is not always desirable. The S3C model avoids this issue by controlling the sparseness (locations of nonzero elements) and the magnitude separately, via the binary variable $\mathbf{b}$ and the continuous variable $\mathbf{s}$ accordingly. Another issue is that those ℓ_1 norm-based sparse coding methods generally do not produce true-sparse coefficient; instead, the majority of the elements are just close to 0, or "approximately" sparse. In contrast, the S3C model utilizes the binary variable $\mathbf{b}$ (the spike) to places a greater restriction on the sparseness of the coefficient $\mathbf{x}$.

---

[1]In Goodfellow et al. [2012], logistic sigmoid function is used for drawing binary latent variable $\mathbf{b}$, however we use Bernoulli distribution here, which doesn't affect the discussion.

The connection of S3C to restricted Boltzmann machine (RBM) is also described in Goodfellow et al. [2012]. Similar ideas have been explored in Zhou et al. [2009] for dictionary learning and shown several advantages over other methods like K-SVD [Aharon et al., 2005].

A survey on recent advances of graphical models in sparse coding was presented in Cevher et al. [2010].

CHAPTER 3

Dictionary Learning Algorithms

In this chapter, we introduce some of the most well-known dictionary learning algorithms in the recent literature, with an emphasis on those reported in the period from 2008–2014. We present these dictionary algorithms by grouping them into different categories, according to their learning outcomes or how they are typically used: learning dictionary for sparse representation (Sec. 3.1), learning dictionary for classification tasks (Sec. 3.2), joint learning of multiple dictionaries (Sec. 3.3), on-line dictionary learning (Sec. 3.4), and statistical dictionary learning (Sec. 3.5).

3.1 RECONSTRUCTIVE DICTIONARY LEARNING

Among the early work explicitly discussing dictionary learning is the MOD algorithm presented in Engan et al. [1999a,b]. The algorithm aims at approximating an input signal $\mathbf{x}$ by a linear combination of frame vectors (i.e., dictionary atoms in this book) $\{\mathbf{d}_1, \mathbf{d}_2, \cdots, \mathbf{d}_K\}$:

$$\bar{\mathbf{x}} = \sum_i \alpha_i \mathbf{d}_i \tag{3.1}$$

under the constraint that only T elements of α is nonzero. To solve Eq. 3.1, the MOD algorithm employs a simple iterative procedure, as summarized in Algorithm 5.

Dictionary learning did not become widely employed until the development of the K-SVD algorithm [Aharon et al., 2005]. K-SVD aims at finding the dictionary, under which the training data can be sparse-represented. Accordingly, the task can be formulated as the following optimization problem:

$$\mathbf{D}, \alpha : \min_{\mathbf{D}, \alpha} \sum_{i=1}^{N} \frac{1}{2} \|\mathbf{x}_i - \mathbf{D}\alpha_i\|_2^2 \text{ s.t., } \|\alpha_i\|_0 \leq T \text{ and } \|\mathbf{d}_i\|_2 = 1 \; \forall i \tag{3.2}$$

where $\mathbf{x}_i \in \mathbb{R}^{d \times 1}$ is the i_{th} training data point (as a column of $\mathbf{X} \in \mathbb{R}^{d \times N}$), $\mathbf{D} \in \mathbb{R}^{d \times K}$ the dictionary, $\alpha_i \in \mathbb{R}^{K \times 1}$ the sparse coefficient for $\mathbf{X}_i$, and T the sparsity constraint. Each column of the dictionary is required to have unit ℓ_2 norm.

According to Eq. 3.2, each training data sample should be reconstructed with at most T columns from $\mathbf{D}$ (the columns of the dictionary will be referred to as dictionary atoms) with

---

Algorithm 5 The MOD algorithm

Input : training data $\mathbf{X}$, number of dictionary atoms K and sparse constraint T
Output : dictionary $\mathbf{D}$ and sparse coefficient α

1: Initialize $\mathbf{D}$
2: **while not** converged **do**
3:      **for** each training data $\mathbf{x}_i$ **do**
4:          Compute the sparse coefficient α_i with orthogonal matching pursuit (OMP)
5:      **end for**
6:      Compute the residuals $\mathbf{R} = \mathbf{X} - \mathbf{D}\alpha$ of sparse reconstruction
7:      Update dictionary as $\mathbf{D} = \mathbf{D} + \delta$ with $\delta = (\alpha\alpha^T)^{-1}\alpha\mathbf{R}$
8:      Normalize the dictionary to have unit ℓ_2 norm for each column
9:      Check stop criterion
10: **end while**

---

minimal error, i.e.,

$$\hat{\mathbf{X}}_i = \mathbf{D}\alpha = \sum_{j:\alpha_i^j \neq 0} \mathbf{d}_j \alpha_i^j. \tag{3.3}$$

K-SVD can be viewed a generalization of the K-means algorithm (and this is also why it is named K-SVD). In K-means, K clusters are generated from the training data and each training sample can be represented by the centroid of one cluster, using the index of the corresponding cluster. Similarly, in K-SVD a dictionary with K dictionary atoms are leaned from the training data and each training sample can be represented by a linear combination of a small subset of the dictionary atoms, according to the sparse coefficient. By relaxing the constraint that each training sample is only related to one cluster center in K-means, K-SVD is able to learn a more compact representation of the data with better reconstruction capability.

A block-coordinate-descent algorithm can be used for solving Eq. 3.2: fix $\mathbf{D}$ to optimize α; then fix α to update $\mathbf{D}$; repeat these two steps until some stop criterion is satisfied. Given $\mathbf{D}$, the sparse coefficient α can be computed by solving the following problem:

$$\alpha : \min_{\alpha} \sum_{i=1}^{N} \frac{1}{2}\|\mathbf{x}_i - \mathbf{D}\alpha_i\|_2^2 \text{ s.t., } \|\alpha_i\|_0 \leq T \; \forall i, \tag{3.4}$$

where each α_i can be computed separately, i.e.,

$$\alpha_i : \min_{\alpha_i} \frac{1}{2}\|\mathbf{x}_i - \mathbf{D}\alpha_i\|_2^2 \text{ s.t., } \|\alpha_i\|_0 \leq T \tag{3.5}$$

which can be solved by many sparse solvers, e.g., orthogonal matching pursuit (OMP).

Given α, $\mathbf{D}$ can be updated by solving the following problems:

$$\mathbf{D} : \min_{\mathbf{D}} \sum_{i=1}^{N} \frac{1}{2} \|\mathbf{x}_i - \mathbf{D}\alpha_i\|_2^2 \text{ s.t., } \|\mathbf{d}_i\|_2 = 1 \ \forall i. \tag{3.6}$$

Considering $\mathbf{x}_i - \mathbf{D}\alpha_i = \mathbf{x}_i - \sum_{k \neq j} \mathbf{d}_k \alpha_i^k - \mathbf{d}_j \alpha_i^j$, $\mathbf{D}$ can be updated by iteratively updating each dictionary atom $\mathbf{d}_j$:

$$\mathbf{d}_j \quad : \quad \min_{\mathbf{d}_j} \sum_{i=1}^{N} \frac{1}{2} \|\mathbf{e}_i - \mathbf{d}_j \alpha_i^j\|_2^2$$

$$\min_{\mathbf{d}_j} \frac{1}{2} \|\mathbf{E} - \mathbf{d}_j \alpha^j\|_F^2$$

$$\text{s.t.,} \quad \|\mathbf{d}_j\|_2 = 1, \tag{3.7}$$

where $\mathbf{e}_i = \mathbf{x}_i - \sum_{k \neq j} \mathbf{d}_k \alpha_i^k$ is the signal residual without dictionary atom $\mathbf{d}_j$ and $\mathbf{E} = [\mathbf{e}_i]_{i=1}^{N} = [\mathbf{e}_1, \mathbf{e}_2, \cdots, \mathbf{e}_n]$. Eq. 3.7 can be solved by singular value decomposition (SVD). By applying SVD, we have $\mathbf{E} = \mathbf{U}\Sigma\mathbf{V}^T$, accordingly

$$\begin{aligned} \mathbf{d}_j &= \mathbf{u}_1 \\ \alpha^j &= \sigma_1 \mathbf{v}_1, \end{aligned} \tag{3.8}$$

where σ_1 is the largest singular value.

While being simple and straightforward, the above method has one drawback: when updating the dictionary atoms, the sparsity of α cannot be preserved. The reason is that, when we make the update $\alpha^j = \sigma_1 \mathbf{v}_1$, some zero-valued elements of α may be set to nonzero values.

To address this issue, another way of updating the dictionary atoms was proposed in Aharon et al. [2005], which is able to preserve the sparsity of the sparse coefficients. In this method, instead of applying SVD to the residual of all signals, only the signals which utilize that dictionary atom will be considered:

$$\mathbf{d}_j \quad : \quad \min_{\mathbf{d}_j} \sum_{i:\alpha_i^j \neq 0} \frac{1}{2} \|\mathbf{e}_i - \mathbf{d}_j \alpha_i^j\|_2^2$$

$$\min_{\mathbf{d}_j} \frac{1}{2} \|\hat{\mathbf{E}} - \mathbf{d}_j \alpha^j\|_F^2$$

$$\text{s.t.,} \quad \|\mathbf{d}_j\|_2 = 1, \tag{3.9}$$

where $\hat{\mathbf{E}} = [\mathbf{e}_i]_{i:\alpha_i^j \neq 0}$. Then by appling SVD to $\hat{\mathbf{E}}$, the dictionary atom $\mathbf{d}_j$ and sparse coefficients $\alpha_{i:\alpha_i \neq 0}^j$ are updated with Eq. 3.8.

In summary, the K-SVD algorithm is described in Algorithm 6. An example of the dictionary learned with the K-SVD algorithm is shown in Fig. 3.1.

Figure 3.1: An examples of the dictionary learned from natural image patches with K-SVD, which contains a total of 256 atoms with each atom of size 8×8 pixels (shown as individual cells).

For the stop criterion, two typical conditions to consider are (1) the algorithm has reached certain maximal number of iterations, or (2) the reconstruction error $\sum_i \|\mathbf{x}_i - \mathbf{D}\alpha\|_2^2$ has become smaller than some pre-specified threshold.

3.1.1   LEARNING SHIFT-INVARIANT DICTIONARIES

An interesting idea was proposed in Aharon and Elad [2008] to learn a dictionary called image signature dictionary (ISD), with the new dictionary itself being a small image from which patches extracted with varying sizes at varying locations are possible atoms for representation. That is, we can write

$$\mathbf{x} = \sum_k \sum_l \alpha_{k,l} \mathcal{C}_{k,l}(\mathbf{D}), \tag{3.10}$$

where $\mathbf{x} \in \mathbb{R}^{n \times 1}$ is an input image, α is the coefficient, $\mathbf{D} \in \mathbb{R}^{\sqrt{m} \times \sqrt{m}}$ is the dictionary (i.e., the ISD), and $\mathcal{C}_{k,l}(\mathbf{D})$ is the patch extraction operator, which extracts a $\sqrt{n} \times \sqrt{n}$ patch at location (k, l) from $\mathbf{D}$ then converts it to an $n \times 1$ vector. All possible (k, l) are simply enumerated in $\mathbf{D}$. As opposed to the dictionary learned from K-SVD, a single ISD can present image patches of varying size, by different patch extraction operators. In addition, the ISD is visually self-explanatory.

Algorithm 6 The K-SVD algorithm

Input : training data $\mathbf{X}$, number of dictionary atoms K and sparse constraint T
Output : dictionary $\mathbf{D}$ and sparse coefficient α

 1: Initialize $\mathbf{D}$
 2: **while not** converged **do**
 3:     **for** each training data $\mathbf{x}_i$ **do**
 4:         Compute the sparse coefficient α_i with OMP
 5:     **end for**
 6:     **for** each dictionary atom $\mathbf{d}_j$ **do**
 7:         Compute the overall reconstruction error $\mathbf{E} = \mathbf{X} - \sum_{k \neq j} \mathbf{d}_k \alpha^k$
 8:         Find the set of signals which utilize $\mathbf{d}_j$, i.e., $\Omega = \{i \,|\, \alpha_i^j \neq 0\}$
 9:         Restrict the error to those signals $\hat{\mathbf{E}} = [\mathbf{e}_i]_{i \in \Omega}$
10:         Apply singular value decomposition (SVD) $\hat{\mathbf{E}} = \mathbf{U} \Sigma \mathbf{V}^T$
11:         Update $\mathbf{d}_j = \mathbf{u}_1$ and $\alpha_\Omega^j = \sigma_1 \mathbf{v}_1$.
12:     **end for**
13:     Check stop criterion
14: **end while**

The ISD can be learned with a modified K-SVD algorithm. In the sparse coding step, a modified OMP algorithm is used to compute the sparse coefficient by fixing the dictionary:

$$\alpha \quad : \quad \min_\alpha \sum_{i=1}^{N} \frac{1}{2} \|\mathbf{x}_i - \sum_k \sum_l \alpha_i^{k,l} \mathcal{C}_{k,l}(\mathbf{D})\|_2^2 \tag{3.11}$$

$$\min_\alpha \sum_{i=1}^{N} \frac{1}{2} \|\mathbf{x}_i - \sum_k \sum_l \alpha_i^{k,l} \mathbf{d}_{k,l}\|_2^2$$

$$\text{s.t.} \quad \|\alpha_i\|_0 \leq T \;\; \forall i.$$

In Eq. 3.11, we simply enumerate the patches from all possible locations (k, l) to build a new dictionary $\{\mathbf{d}_{k,l}\}$. This appears to be very costly, since we need to consider all possible patches. However, utilizing the inner structure of the ISD, the computation of the projections of the signal to all dictionary atoms $\{\mathbf{d}_{k,l}\}$ (or all possible locations of $\mathbf{D}$) can be achieved with convolution efficiently. With sparse coefficient α computed, the dictionary $\mathbf{D}$ can be computed by gradient descent, which is detailed in Sec. 3.3 and 3.4 of Aharon and Elad [2008]. A similar algorithm was proposed in Mailhé et al. [2008] for learning a shift-invariant dictionary for a very long 1D signal.

The idea of learning shift-invariant dictionary was further studied in Bristow et al. [2013], where a convolutional dictionary is learned from the training data. The objective function of Bristow et al. [2013] and the corresponding optimization problem are given as:

$$\mathbf{D}, \alpha : \min_{\mathbf{D},\alpha} \sum_{i=1}^{N} \frac{1}{2} \|\mathbf{x}_i - \sum_j \mathbf{d}_j * \alpha_i^j\|_2^2 \text{ s.t., } \|\alpha_i\|_0 \leq T \text{ and } \|\mathbf{d}_i\|_2 = 1 \ \forall i, \qquad (3.12)$$

where each training sample $\mathbf{x}_i$ can be either 1D signals like audio or 2D signals like images and $\mathbf{d}_i$ can be either a 1D vector or a 2D matrix accordingly. By utilizing the fact that, convolution in the spatial domain is equivalent to element-wise multiplication in the frequency domain, the problem is solved in the frequency domain by applying Lagrange multipliers: [1]

$$\hat{\mathbf{D}}, \alpha, \beta : \min_{\hat{\mathbf{D}},\alpha,\beta} \sum_{i=1}^{N} \frac{1}{2} \|\hat{\mathbf{x}}_i - \sum_j \hat{\mathbf{d}}_j \bullet \beta_i^j\|_2^2 \text{ s.t., } \|\alpha_i\|_0 \leq T \text{ and } \|\mathbf{d}_i\|_2 = 1 \text{ and } \beta = \mathbf{F}\alpha \ \forall i,$$

$$(3.13)$$

where $\beta = \mathbf{F}\alpha$ is the Fourier coefficient and $\mathbf{F}$ is the dictionary for DFT as described in Sec. 1.1. According to Parseval's Theorem, $\|\hat{\mathbf{d}}_j\|_2 = 1 \equiv \|\mathbf{d}_j\|_2 = 1$. Fig. 3.2 illustrates an example from Bristow et al. [2013].

Signals from physical systems may often be modelled by a clean source convoluted with some kernel. A typical example is a low-resolution imaging system capturing an image that may be viewed as a low-pass-filtered version of an underlying, "true" high-resolution image. The convolution process may introduce artifacts that can cause problems for sparse recovery of such signals. Considering this fact, Barchiesi and Plumbley [2011] proposed to learn a dictionary for convoluted signals with the constraints of minimal reconstruction error and sparse coefficients. Consider a signal $\mathbf{x} \in \mathbb{R}^{N \times 1}$, which is sparsely represented by a dictionary $\mathbf{D} \in \mathbb{R}^{N \times K}$ as $\mathbf{x} = \mathbf{D}\alpha$. Suppose that we do not have direct observation of $\mathbf{x}$, but instead its convoluted version $\mathbf{y} \in \mathbb{R}^{(N-L+1) \times 1}$:

$$y_i = \sum_{j=0}^{L-1} h_l x_{i-l} \rightarrow \mathbf{y} = \mathbf{h} * \mathbf{x} = \mathbf{h} * (\mathbf{D}\alpha), \qquad (3.14)$$

where $\mathbf{h} \in \mathbb{R}^{L \times 1}$ is the convolution kernel and $*$ is the convolution operator. The convolution operation can be converted to matrix product by utilizing the Toeplitz convolutive matrix $\mathbf{H}$:

$$\mathbf{y} = \mathbf{H}\mathbf{x} = \mathbf{H}\mathbf{D}\alpha. \qquad (3.15)$$

In many applications, the convolution kernel or convolutive matrix cannot be known precisely, which may also be signal-dependent. We can still learn the dictionary $\mathbf{D}$ by assuming that $\mathbf{H}$ is an identity matrix (i.e., $\mathbf{h}$ is a Dirac function), or absorbing $\mathbf{H}$ into $\mathbf{D}$, as done in many

---

[1]We think the reformulation in Bristow et al. [2013] (Eq. 2 to Eq. 3) is less accurate than the one described here. A signal is sparse does not mean its Fourier coefficient is also sparse.

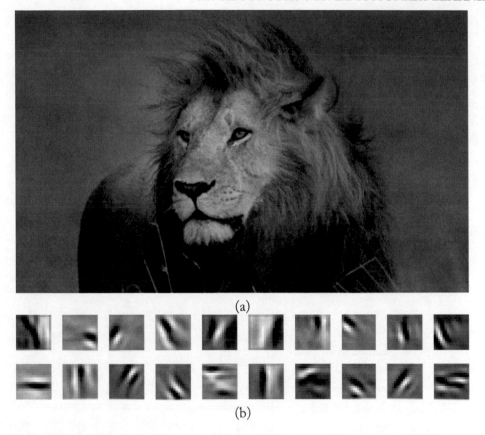

Figure 3.2: A sample dictionary (b) (only some of the atoms shown) learned from the image in (a). The atoms include expression of generic Gabor-like filters as well as specific filters like highlighted "eyes". The example is from Fig. 1 of Bristow et al. [2013].

existing dictionary algorithms. However, such learned dictionaries will deviate from the ground truth (if there is one) and thus reduce its performance on sparse coding, as analyzed in Sec. 2.2 of [Barchiesi and Plumbley, 2011]. Assume that $\mathbf{F}$ is the dictionary for DFT as described in Sec. 1.1 and $\mathbf{F}^H$ is its conjugate transpose. We know $\mathbf{F}^H$ is orthogonal, then we have:

$$\|\mathbf{Y} - \mathbf{h} * (\mathbf{D}\alpha)\|_F^2 = \|\mathbf{F}^H\mathbf{Y} - \mathbf{F}^H\mathbf{h} * (\mathbf{D}\alpha)\|_F^2. \tag{3.16}$$

By utilizing the fact that, convolution[2] in the frequency domain is equivalent to element-wise multiplication in the spatial domain, we have:

$$\|\mathbf{F}^H\mathbf{Y} - \mathbf{F}^H\mathbf{h} * (\mathbf{D}\alpha)\|_F^2 = \|\hat{\mathbf{Y}} - \mathcal{D}(\hat{\mathbf{h}})\hat{\mathbf{D}}\alpha\|_F^2, \tag{3.17}$$

[2]More precisely, the circular convolution. Accordingly, we need to do zero-padding during Fourier Transform, which is detailed in Sec. 3.1.1 of Barchiesi and Plumbley [2011].

Algorithm 7 The algorithm for dictionary learning of convolved signal

---

Input : training data $\mathbf{Y}$, number of dictionary atoms K and sparse constraint T
Output : dictionary $\mathbf{D}$, convolution kernel $\mathbf{h}$ and sparse coefficient α

1: Initialize $\mathbf{D}$ and $\mathbf{h}$
2: Convert the $\mathbf{D}$ and $\mathbf{Y}$ to frequency domain: $\hat{\mathbf{D}}$ and $\hat{\mathbf{Y}}$.
3: **while not** converged **do**
4:     Compute the sparse coefficient α via OMP
5:     Update dictionary in frequency domain $\hat{\mathbf{D}}$ and α via K-SVD
6:     Compute the convolution kernel in frequency domain $\hat{\mathbf{h}}$ with Eq. 3.18
7:     Check stop criterion
8: **end while**
9: Convert $\hat{\mathbf{D}}$ and $\hat{\mathbf{h}}$ back to spatial domain: $\mathbf{D}$ and $\mathbf{h}$

---

where $\hat{\cdot}$ is the Fourier coefficient of the input, accordingly $\hat{\mathbf{D}}$ is the dictionary in the frequency domain and $\mathcal{D}(\cdot)$ is the diagonal operator.

The objective function can be finally written as:

$$\alpha, \tilde{\mathbf{h}}, \hat{\mathbf{D}} : \min_{\alpha, \tilde{\mathbf{h}}, \hat{\mathbf{D}}} \frac{1}{2} \|\hat{\mathbf{Y}} - \tilde{\mathbf{h}}\hat{\mathbf{D}}\alpha\|_F^2 \text{ s.t. } \|\alpha_i\|_0 \leq T \ , \ \|\hat{\mathbf{d}}_j\|_2 = 1 \ \forall i, j, \tag{3.18}$$

where we use $\tilde{\mathbf{h}}$ for $\mathcal{D}(\hat{\mathbf{h}})$. Equation 3.18 can be solved via block coordinate descent. By converting the input to the frequency domain and fixing $\tilde{\mathbf{h}}$, the sparse coefficient α and dictionary $\hat{\mathbf{D}}$ can be learned via the K-SVD algorithm.

To summarize, the algorithm of Barchiesi and Plumbley [2011] is described in Algorithm 7.

Proposed in Kavukcuoglu et al. [2010] was another solution of convolutional dictionary learning, where a trainable, feed-forward, non-linear encoder module was used to produce a fast approximate of the sparse code. The objective function can be written as:

$$\mathbf{D}, \alpha : \min_{\mathbf{D}, \alpha} \sum_{i=1}^{N} \frac{1}{2} \|\mathbf{x}_i - \sum_j \mathbf{d}_j * \alpha_i^j\|_2^2 + \|\alpha_i - f(\mathbf{w}^i \mathbf{x}_i)\|_2^2 + \lambda \|\alpha_i\|_1, \tag{3.19}$$

where $f(\mathbf{w}^i \mathbf{x}_i)$ is a point-wise nonlinear function parameterized by $\mathbf{w}^i$ to provide fast approximate to the sparse code α_i. The choice of such nonlinear function $f()$ and the optimization algorithm was discussed in Kavukcuoglu et al. [2010], which interested readers can refer to for more details.

3.1.2   LEARNING DICTIONARIES IN THE KERNEL SPACE

The kernel trick can be applied in dictionary learning, which makes the dictionary capable of capturing complex intrinsic structures of the data as well as enables learning in high-dimensional spaces. In Nguyen et al. [2012] and Van Nguyen et al. [2013], a kernel extension of the K-SVD algorithm was proposed to enable dictionary learning in a high-dimensional space. In this approach, the objective function is written as:

$$\mathbf{D}, \alpha : \min_{\mathbf{D},\alpha} \sum_{i=1}^{N} \frac{1}{2} \|\Phi(\mathbf{x}_i) - \Phi(\mathbf{D})\alpha_i\|_2^2 \text{ s.t., } \|\alpha_i\|_0 \leq T \text{ and } \|\Phi(\mathbf{d}_i)\|_2 = 1 \ \forall i, \qquad (3.20)$$

where $\Phi(\cdot)$ is the kernel function. By assuming that the dictionary atoms lie within the subspace spanned by the training data, we can write $\Phi(\mathbf{X})\beta$. Accordingly, the problem can be written as:

$$\beta, \alpha : \min_{\beta,\alpha} \sum_{i=1}^{N} \frac{1}{2} \|\Phi(\mathbf{x}_i) - \Phi(\mathbf{X})\beta\alpha_i\|_2^2 \text{ s.t., } \|\alpha_i\|_0 \leq T \ \forall i, \qquad (3.21)$$

where the constraint $\|\Phi(\mathbf{d}_i)\|_2 = 1$ is removed. By assuming a Mercer kernel, we have:

$$\|\Phi(\mathbf{x}_i) - \Phi(\mathbf{X})\beta\alpha_i\|_2^2 \equiv \text{tr}((\mathbf{I} - \beta\alpha)^T \mathbb{K}(\mathbf{X}, \mathbf{X})(\mathbf{I} - \beta\alpha), \qquad (3.22)$$

where the kernel $\mathbb{K}(\mathbf{X}, \mathbf{X})$ is a positive semi-definite matrix, with $\mathbb{K}(\mathbf{X}, \mathbf{X})_{i,j} = \langle \Phi(\mathbf{x}_i), \Phi(\mathbf{x}_j) \rangle$.

Equation 3.22 can be solved similarly as done in the K-SVD algorithm, i.e., through alternately updating the sparse coefficient and the dictionary. The OMP and K-SVD algorithms, however, need to be extended to the kernel space accordingly. In Kernel OMP (KOMP), given a signal $\mathbf{z}$, the residual $\mathbf{r}_s$ after dictionary atoms $\{\mathbf{d}_j | j \in I_s\}$ selected can be computed as:

$$\mathbf{r}_s^T (\Phi(\mathbf{X})\beta_i) = (\Phi(\mathbf{z}) - \Phi(\mathbf{z}_s))^T \Phi(\mathbf{X})\beta_i = (\mathbb{K}(\mathbf{z}, \mathbf{X}) - \mathbb{K}(\mathbf{X}, \mathbf{X})\mathbf{z}_s^T)\beta_i, \qquad (3.23)$$

where $\Phi(\mathbf{z}_s) = \Phi(\mathbf{X})\beta\alpha_s$ and α_s is the current estimation of the sparse coefficients. The next atom is picked such that $\|\mathbf{r}_s^T (\Phi(\mathbf{X})\beta_i)\|$ is maximal. In Kernel K-SVD, the reconstruction error for updating the k_{th} atom is written as:

$$\|\Phi(\mathbf{X}) - \Phi(\mathbf{X})\beta\alpha\| = \|\Phi(\mathbf{X})(\mathbf{I} - \sum_{j \neq k} \beta_j \alpha^j) - \Phi(\mathbf{X})\beta_k \alpha^k\|. \qquad (3.24)$$

SVD is then applied on $\Phi(\mathbf{X})(\mathbf{I} - \sum_{j \neq k} \beta_j \alpha^j)$ to update β_k and α^k.

Similar idea was found in Xie and Feng [2009], which utilizes kernel fuzzy codebook estimation and gradient descent. In Harandi et al. [2012], the dictionary is required to be a positive semi-definite matrix and Bregmann matrix divergence is used as the kernel function. The log-Euclidean kernel was used in Li et al. [2013] to produce a complete inner-product space.

3.1.3   OTHER DICTIONARY LEARNING ALGORITHMS

In this subsection, we introduce additional dictionary learning approaches not covered in the previous subsections. In Labusch et al. [2009], a Sparse Coding Neural Gas algorithm was proposed, which is based on the combination of the original Neural Gas (NG) algorithm and Oja's rule [Oja, 1982]. In the NG algorithm, hard-competitive learning is replaced by soft-competitive learning, which controls the learning rate of the dictionary atoms according to the sequential order of their distances to the current training data. Consider the following sorted sequence of distances:

$$- (\mathbf{d}_{l_1}^T \mathbf{x}) \leq \cdots \leq -(\mathbf{d}_{l_k}^T \mathbf{x}) \leq \cdots \leq -(\mathbf{d}_{l_K}^T \mathbf{x}). \tag{3.25}$$

Then with the Oja's rule, in the t_{th} iteration, the dictionary atom can be updated as:

$$\Delta \mathbf{d}_{l_k} = \alpha_t e^{\frac{k}{\lambda_t}} \mathbf{y}(\mathbf{x} - \mathbf{y} \mathbf{d}_{l_k}), \tag{3.26}$$

where λ_t is some constant at time t, $\mathbf{y}_j = \mathbf{d}_j^T \mathbf{x}$ and $\alpha_t = \alpha_0 (\frac{\alpha_T}{\alpha_0})^{\frac{t}{T}}$ is the learning rate. Compared with the K-SVD algorithm, the Sparse Coding Neural Gas algorithm does not require singular value decomposition in each iteration of updating each dictionary atom, which is required in K-SVD. As a result, it is much more efficient than K-SVD. Experiments also showed that the Sparse Coding Neural Gas algorithm is more robust to additive noise than K-SVD.

In K-SVD, the ℓ_0 norm is used for measuring sparsity, which, being nonconvex and noncontinuous, makes the optimization difficult. As a result, several alternatives of the ℓ_0 norm has been proposed in different algorithms. In Zayyani and Babaie-Zadeh [2009], the ℓ_0 norm was replaced by a convex and continuous measurement, which is named smoothed-ℓ_0:

$$\|\mathbf{x}\|_0 \rightarrow m - \sum_{i=1}^{m} e^{-\frac{x_i^2}{2\sigma^2}}, \tag{3.27}$$

where $\mathbf{x} \in \mathbb{R}^{m \times 1}$ and σ is a small constant. Accordingly, updating the sparse coefficients and the dictionary atoms can be achieved via soft-thresholding, resulting in much less computation than required by K-SVD. In Jafari and Plumbley [2011], the ℓ_0 norm was replaced by the ratio of the ℓ_1 norm and the ℓ_2 norm:

$$\|\mathbf{x}\|_0 \rightarrow \frac{\|\mathbf{x}\|_1}{\|\mathbf{x}\|_2} = \frac{\sum_i \|x_i\|}{\sum_i x_i^2} \tag{3.28}$$

which is called sparsity index. Accordingly, a greedy adaptive dictionary algorithm (GAD) was proposed. GAD was evaluated on speech representation and speech denoising, and improved performances were reported. In many other algorithms, the ℓ_0 norm is simply replaced by the ℓ_1, which is differentiable and continuous. In Bao et al. [2013b], an alternating proximal method based algorithm was proposed for the problem of dictionary learning. The algorithm alternately solves two proximal problems to update the dictionary and sparse coefficients accordingly, via gradient descent. The method was proven to have global convergence with a sub-linear convergence rate, which is not found in many other papers. A direct optimization method was proposed in Rakotomamonjy [2013], which avoids the usual techniques involving alternating the updating of the sparse coefficients and the dictionary atoms. More specifically, it performs a joint proximal gradient descent step over both the sparse coefficients and the dictionary atoms, where in the k_{th} iteration the following updates are made:

$$\mathbf{D}^{t+1} = \mathcal{P}(\mathbf{D}^t + \eta_t(\mathbf{X} - \mathbf{D}^t\alpha^t)(\alpha^t)^T) \tag{3.29}$$

$$\alpha^{t+1} = \mathcal{S}^{\mu}_{\eta\lambda}(\alpha^t + \eta_t(\mathbf{D}^t)^T(\mathbf{X} - \mathbf{D}^t\alpha^t)), \tag{3.30}$$

where μ, η, λ are the parameters of the algorithm, the projection operator $\mathcal{P}$ normalizes each column of $\mathbf{D}$ to unit ℓ_2 norm and $\mathcal{S}^{\mu}_{\eta}$ is the clipped (to $-\mu \sim \mu$) soft-threshold (at η) operator.

In Yaghoobi et al. [2009a], the majorization method is used for dictionary learning, which substitutes the original objective function with a surrogate function that is updated in each step. For example, by applying majorization,

$$\|\mathbf{X} - \mathbf{D}\alpha\|_F^2 \leq \|\mathbf{X} - \mathbf{D}\alpha\|_F^2 + c\|\alpha - \hat{\alpha}\|_F^2 - \|\mathbf{D}\alpha - \mathbf{D}\hat{\alpha}\|_F^2 \tag{3.31}$$

with $c \geq \|\mathbf{D}^T\mathbf{D}\|_F$ a constant. The major benefit of the majorization method is that, it is flexible to incorporate different types of constraints.

In sparse coding, a small dictionary is desired, as the size of the dictionary is influential on computational efficiency and the optimality for greedy strategies (such as matching pursuit) that are often employed. Accordingly, several extensions of K-SVD have been proposed to optimize the size of the learned dictionary. E-KSVD [Mazhar and Gader, 2008] is such an example that is able to learn a dictionary of optimal size without compromising its sparse reconstruction performance. This is achieved by combing competitive agglomeration (CA) with K-SVD. CA aims to discover the optimal number of clusters during the clustering process. E-KSVD starts with a dictionary with a large number of dictionary atoms, which can be learned with standard K-SVD, and gradually prunes off the under-utilized dictionary atoms or atoms with high similarity to other atoms (e.g., via correlation analysis). This results in a dictionary with much-reduced redundance among the atoms.

In Yaghoobi et al. [2009b], parsimony dictionary learning was proposed, which aims to learn a dictionary with minimal size and minimal reconstruction error, under the sparsity constraint. In the objective function, the size of the dictionary is penalized by a $\ell_{1,q}$ mixture norm on the dictionary. For optimization, a majorization-based method is proposed. Experiments on audio signal encoding were reported to show that the method improves the rate-distortion performance compared with other dictionary learning methods. In Jiang et al. [2012], a method of selecting dictionary atoms from some known set of basis was proposed. The method greedily selects atoms from a known set of bases by maximizing the reduction of the variance of the training data. It was shown that, if the atoms of the known set of bases are incoherent enough, the problem becomes sub-modular. Accordingly, the proposed greedy method is guaranteed to provide a polynomial approximate solution to the global optimum.

In some applications, dictionaries with some special structure or properties may be desired. Several algorithms have been proposed in the literature for this purpose. In Ataee et al. [2010] and

Yang and Zhang [2010], algorithms for learning dictionaries with a structure similar to Gabor functions were proposed. Their simulation experiments show that such Gabor-like dictionaries yield better representation and discriminative ability, than the dictionaries learned by traditional algorithms such as K-SVD. To learn such a dictionary, each dictionary atom is parameterized by scale and frequency of the Gabor function and a steepest descent algorithm is proposed to learn those parameters.

In Rubinstein et al. [2010], the dictionary was assumed to be also sparse-represented by other dictionaries, i.e., $\mathbf{D} = \Phi\Psi$, where Φ is called the base dictionary and Ψ is the sparse coefficient for the sparse dictionary. Such a sparse dictionary is more efficient and more compact to store and transmit than its dense counterparts. In addition, it reduces overfitting and instability in the presence of noise. The problem of learning sparse dictionary can be formulated as:

$$\alpha, \Psi : \min_{\alpha, \Psi} \|\mathbf{X} - \Phi\Psi\alpha\|_F^2 \text{ s.t. } \|\psi_i\|_0 \leq t, \|\alpha_j\|_0 \leq p, \|\Phi\psi_i\|_2 = 1. \tag{3.32}$$

The problem can be solved via block coordinate descent. In particular, the problem of updating one dictionary atom can be written as:

$$\Psi : \min_{\Phi, \Psi} \|\mathbf{X} - \Phi\psi_i\alpha^i\|_F^2 \text{ s.t. } \|\psi_i\|_0 \leq t, \|\Phi\psi_i\|_2 = 1, \tag{3.33}$$

where the base dictionary Φ is assumed to be known, e.g., the DCT matrix. This problem is closely related to another problem named Sparse Matrix Approximation (SMA) raised in kernel support vector machine [Smola and Schökopf, 2000], and interested readers should check [Rubinstein et al., 2010, Smola and Schökopf, 2000] for more details of the algorithm.

In Sivalingam et al. [2011], learning a dictionary that is a positive definite matrix was studied, where both the training data and the learned dictionary are positive semi-definite. Accordingly, the squared loss is replaced by logarithm-determinant. Several optimization algorithms were proposed to solve the proposed problem, including gradient descent and online learning. An algorithm was proposed in Esser et al. [2012] for learning nonnegative dictionaries. A dictionary learning algorithm was proposed in Yao et al. [2011] such that both the dictionary and the codes are sparse. The learned dictionary is applied to action attributes and parts for action recognition from still images.

Incoherent dictionaries are more appropriate for sparse approximation. The coherence of a dictionary can be measured as:

$$\mu_{\mathbf{D}} = \max_{i \neq j} \frac{|\langle \mathbf{d}_i, \mathbf{d}_j \rangle|}{\|\mathbf{d}_i\|_2 \|\mathbf{d}_j\|_2}. \tag{3.34}$$

If the columns of the dictionary are normalized to unit ℓ_2 norm, then $\mu_{\mathbf{D}} = \max_{i \neq j} \|\langle \mathbf{d}_i, \mathbf{d}_j \rangle\|$ and obviously $0 \leq \mu_{\mathbf{D}} \leq 1$. For a dictionary that is orthogonal matrix, its coherence is zero. In Bao et al. [2013a], a dictionary learning algorithm was proposed, where the coherence of the learned dictionary is minimized. The resultant dictionary is orthogonal, which is more compact than that learned by other methods. Reducing the coherence of the learned dictionary was also studied in Yaghoobi et al. [2009c], where the coherence of the learned dictionary is measured as the difference of the gram matrix of the dictionary ($g_{\mathbf{D}} = \mathbf{D}^T \mathbf{D}$) to the identity matrix. Experiments show that the dictionary learned by this method has some advantages over dictionary learned by other methods, in the sense that fewer dictionary atoms are required to achieve the same reconstruction error.

In compressive sensing, it has been shown that sparse signals (θ) can be recovered from only a few samples ($\mathbf{y}$), which is obtained by projecting the signals with certain sampling matrix (Φ)

$$\mathbf{y} = \Phi \mathbf{x} = \Phi \Psi \theta. \tag{3.35}$$

Compressive sensing requires that the sampling matrix Φ and the dictionary matrix Ψ should be as incoherent (orthogonal) as possible. The coherence of Φ and Ψ is measured as:

$$\mu_{\Phi,\Psi} = \max_{i,j} \frac{|\langle \Phi_i, \Psi_j \rangle|}{\|\Phi_i\|_2 \|\Psi_j\|_2}. \tag{3.36}$$

Several types of sampling matrices have been proposed, e.g., matrices drawn from Gaussian distributions, matrices with columns randomly selected from Fourier basis. Many efforts have been spent on algorithms of recovering the sparse signals and learning dictionary adaptively from the data; but much fewer efforts have been found in learning data-adaptive sampling matrices. A rare example is in Duarte-Carvajalino and Sapiro [2009], where an algorithm of learning the dictionary and sampling matrix simultaneously was proposed. In Duarte-Carvajalino and Sapiro [2009], given the dictionary, which is learned via K-SVD, the incoherence constraint of the sampling matrix is formulated as requiring the gram matrix ($\Psi^T \Phi^T \Phi \Psi$) to be as close to the identity matrix as possible. The eigen decomposition is used to find the optimal sampling matrix under such a constraint. Experiments showed that the method achieves better results than using a non-adaptive random sampling matrix or learning a sampling matrix independently from the dictionary, in image reconstruction tasks.

In some settings such as image classification or dimension reduction, the sparse coefficient for similar signals should be as similar as possible. However, dictionary learned by algorithms like K-SVD cannot guarantee that. This problem was addressed in Mairal et al. [2009b] by combining group sparse coding with dictionary learning. Signals of each group, which are obtained via K-nearest neighbors (KNN), are sparse coded under sparsity constraint. The $\ell_{p,q}$ mixture norm is used for the group sparsity constraint, where the typical choice for (p, q) is $(1, 2)$ and $(0, \infty)$. Experiments on image denoising and image demosaicking showed effectiveness of the method. This problem was also studied in Gao et al. [2010], Zheng et al. [2011], and Zhou and Barner [2013], where the Laplacian matrix was used to enforce the constraint that similar input signals would be coded based on similar sparse coefficients, resulting in:

$$\min_{\alpha} \sum_{i,j} d(\mathbf{x}_i, \mathbf{x}_j) \|\alpha_i - \alpha_j\|_2^2 \rightarrow \min_{\alpha} \alpha^T \mathbf{L} \alpha. \tag{3.37}$$

A tensor dictionary learning algorithm was proposed in Peng et al. [2013] for denoising multi-spectral images. The algorithm utilizes both the spatial-redundancy, such that the 3D patches can be sparse represented by the dictionary, and the spectral redundancy, where the dictionary is required to be low-rank decomposable. In addition, group sparsity is explored, where the 3D patches are first clustered into multiple groups and group sparsity is enforced so that patches of each group would take the same set of supports.

A dictionary learning algorithms for finding representative items from the training data was proposed in Elhamifar et al. [2012]. Similar to K-medoids [Kaufman and Rousseeuw, 1987], the atoms of the dictionary are data from the training data:

$$\mathbf{d} = \mathbf{X}\mathbf{c} \text{ s.t. } \|\mathbf{c}\|_0 = 1, 1^T \mathbf{c} = 1, \tag{3.38}$$

where each dictionary atom $\mathbf{d}$ is one of the data points in the training set $\mathbf{X}$. This problem is very useful in applications like video summarization. The problem can be formulated as:

$$\mathbf{D}, \alpha \quad : \quad \min_{\mathbf{C},\alpha} \frac{1}{2}\|\mathbf{X} - \mathbf{D}\alpha\|_F^2 \text{ s.t. } \|\alpha_i\|_0 \leq T, 1^T \alpha_i = 1 \; \forall i \tag{3.39}$$

$$\beta \quad : \quad \min_{\beta} \frac{1}{2}\|\mathbf{X} - \mathbf{X}\beta\|_F^2 \text{ s.t. } \|\beta_i\|_0 \leq T, 1^T \beta_i = 1 \; \forall i \tag{3.40}$$

where $\beta = \mathbf{C}\alpha$. The sparse codes under such a dictionary are invariant over a global translation of the data. An analysis of the relationship of the learned dictionary atoms to the convex hull of the data was also presented.

3.2 DISCRIMINATIVE DICTIONARY LEARNING

K-SVD learns a dictionary for reconstruction with sparsity constraint. However, in classification tasks, such a dictionary may not be optimal. To this end, many variants have been proposed, which aims at introducing some discriminative power to the learned dictionary. Depending on the way of introducing the discriminative power, we divide these variants into the following categories: explicit discriminative dictionary learning and implicit discriminative dictionary learning.

3.2.1 EXPLICIT DISCRIMINATIVE DICTIONARY LEARNING

In explicit discriminative dictionary learning, typically the objective function includes some classification error term explicitly. For example, in Pham and Venkatesh [2008], given training data $\mathbf{X}$ and corresponding labels $\mathbf{Y}$, the objective function can be written as[3]

$$\mathbf{D}, \alpha, \mathbf{W} \quad : \quad \min_{\mathbf{D},\alpha,\mathbf{W}} \|\mathbf{Y} - \mathbf{W}^T\alpha\|_F^2 + \rho\|\mathbf{X} - \mathbf{D}^T\alpha\|_F^2 + \gamma\|\mathbf{W}\|_F^2 \tag{3.41}$$
$$\text{s.t.} \quad \|\alpha_i\|_0 \leq \epsilon, \|\mathbf{d}_i\|_2^2 = 1,$$

---

[3]The reconstruction error of the unlabeled data is ignored here.

where $\mathbf{D}$ is the dictionary, $\mathbf{W}$ is the classifier, and α is the sparse coefficient. The discriminative power of the dictionary is explicitly represented by the classifier $\mathbf{W}$.

To solve Eq. 3.41, a block coordinate descent algorithm is used, where the dictionary, classifier and sparse coefficient are updated alternately. Given $\mathbf{D}$ and α, the classifier can be updated via solving the following problem:

$$\mathbf{W} : \min_{\mathbf{W}} \|\mathbf{Y} - \mathbf{W}^T \alpha\|_F^2 + \gamma \|\mathbf{W}\|_F^2 \tag{3.42}$$

which is a least square problem and the solution is $\mathbf{W} = (\alpha\alpha^T + \gamma\mathbf{I})^{-1}\alpha\mathbf{Y}^T$.

Given $\mathbf{W}$, the updating of the sparse coefficient α and dictionary $\mathbf{D}$ is related to the following problem:

$$\mathbf{D}, \alpha \quad : \quad \min_{\mathbf{D},\alpha} \|\mathbf{Y} - \mathbf{W}^T \alpha\|_F^2 + \rho\|\mathbf{X} - \mathbf{D}^T \alpha\|_F^2 \tag{3.43}$$

$$\min_{\mathbf{D},\alpha} \left\| \begin{bmatrix} \mathbf{Y} \\ \sqrt{\rho}\mathbf{X} \end{bmatrix} - \begin{bmatrix} \mathbf{W} \\ \sqrt{\rho}\mathbf{D} \end{bmatrix} \alpha \right\|_F^2$$

$$\text{s.t.} \quad \|\alpha_i\|_0 \leq \epsilon, \ \|\mathbf{d}_i\|_2^2 = 1.$$

To maintain the sparsity of the coefficient, the dictionary is updated atom by atom. For atom $\mathbf{d}_i$, we only consider the data whose sparse coefficient is nonzero for $\mathbf{d}_i$:

$$\mathbf{d}_i, \alpha^i \quad : \quad \min_{\mathbf{d}_i,\alpha^i} \left\| \begin{bmatrix} \mathbf{Y} \\ \sqrt{\rho}\mathbf{X} \end{bmatrix} - \sum_{j \neq i} \begin{bmatrix} \mathbf{w}_j \\ \sqrt{\rho}\mathbf{d}_j \end{bmatrix} \alpha^j \right\|_F^2 - \left\| \begin{bmatrix} \mathbf{w}_i \\ \sqrt{\rho}\mathbf{d}_i \end{bmatrix} \alpha^i \right\|_F^2 \tag{3.44}$$

$$\min_{\mathbf{d}_i,\alpha^i} \left\| \begin{bmatrix} \hat{\mathbf{Y}} \\ \sqrt{\rho}\hat{\mathbf{X}} \end{bmatrix} - \begin{bmatrix} \mathbf{w}_i \\ \sqrt{\rho}\mathbf{d}_i \end{bmatrix} \alpha^i \right\|_F^2$$

$$\min_{\mathbf{d}_i,\alpha^i} \left\| \begin{bmatrix} \hat{\mathbf{Y}}_\Omega \\ \sqrt{\rho}\hat{\mathbf{X}}_\Omega \end{bmatrix} - \begin{bmatrix} \mathbf{w}_i \\ \sqrt{\rho}\mathbf{d}_i \end{bmatrix} \alpha^i_\Omega \right\|_F^2$$

$$\text{s.t.} \quad \|\mathbf{d}_i\|_2^2 = 1,$$

where $\begin{bmatrix} \hat{\mathbf{Y}} \\ \sqrt{\rho}\hat{\mathbf{X}} \end{bmatrix} = \begin{bmatrix} \mathbf{Y} \\ \sqrt{\rho}\mathbf{X} \end{bmatrix} - \sum_{j \neq i} \begin{bmatrix} \mathbf{w}_j \\ \sqrt{\rho}\mathbf{d}_j \end{bmatrix} \alpha^j\|_F^2$ and Ω is the set of j in which the coefficient $\alpha^i_j \neq 0$. Equation 3.44 is then solved by solving $\mathbf{d}_i$ and α^i_Ω alternately:

$$\alpha^i_\Omega = \frac{\begin{bmatrix} \mathbf{w}_i \\ \sqrt{\rho}\mathbf{d}_i \end{bmatrix}^T \begin{bmatrix} \hat{\mathbf{Y}}_\Omega \\ \sqrt{\rho}\hat{\mathbf{X}}_\Omega \end{bmatrix}}{\left\| \begin{bmatrix} \mathbf{w}_i \\ \sqrt{\rho}\mathbf{d}_i \end{bmatrix} \right\|_2^2} \tag{3.45}$$

$$\mathbf{d}_i = \frac{\hat{\mathbf{X}}_\Omega (\alpha^i_\Omega)^T}{\|\hat{\mathbf{X}}_\Omega (\alpha^i_\Omega)^T\|_2}. \tag{3.46}$$

Algorithm 8 The algorithm proposed in Pham and Venkatesh [2008]

Input : training data $\mathbf{X}$, label of training data $\mathbf{Y}$ and parameter ρ and ϵ
Output : dictionary $\mathbf{D}$, classifier $\mathbf{W}$ and sparse coefficient α

1: Initialize $\mathbf{D}$ and α with the K-SVD algorithm
2: **while not** converged **do**
3:    Fix $\mathbf{D}$ and α and update classifier as $\mathbf{W} = (\alpha\alpha^T + \gamma\mathbf{I})^{-1}\alpha\mathbf{Y}^T$
4:    Fix $\mathbf{W}$
5:    **for** each dictionary atom $\mathbf{d}_i$ **do**
6:       Identify the set Ω, where the coefficient $\alpha_j : j \in \Omega$ is nonzero for $\mathbf{d}_i$
7:       Update the coefficient α^i_Ω with Eq. 3.45
8:       Update the atom $\mathbf{d}_i$ with Eq. 3.46
9:    **end for**
10:   Check stop criterion
11: **end while**

The algorithm is also presented in Algorithm 8. The algorithm terminates when the maximal number of iterations has been reached or the objective function becomes smaller than certain value.

In Algorithm 8, the classifier $\mathbf{W}$, dictionary $\mathbf{D}$ and coefficient α are updated alternately. One potential drawback is that the convergence could be slow and the procedure is vulnerable to local optimum. In an attempt to overcome these drawbacks, an algorithm termed discriminative K-SVD (D-KSVD) was proposed in Zhang and Li [2010a].

In D-KSVD, the the classifier $\mathbf{W}$, dictionary $\mathbf{D}$ and coefficient α are updated jointly. Compared with Pham and Venkatesh [2008], D-KSVD drops the regularization term $\|\mathbf{W}\|_F$, and has the objective function written as:

$$\mathbf{D}, \alpha, \mathbf{W} \quad : \quad \min_{\mathbf{D},\alpha,\mathbf{W}} \|\mathbf{Y} - \mathbf{W}^T\alpha\|_F^2 + \rho\|\mathbf{X} - \mathbf{D}^T\alpha\|_F^2 \qquad (3.47)$$

$$\min_{\mathbf{D},\alpha,\mathbf{W}} \left\| \begin{bmatrix} \mathbf{Y} \\ \sqrt{\rho}\mathbf{X} \end{bmatrix} - \begin{bmatrix} \mathbf{W} \\ \sqrt{\rho}\mathbf{D} \end{bmatrix} \alpha \right\|_F^2$$

$$\text{s.t.} \quad \|\alpha_i\|_0 \leq \epsilon \,, \|\mathbf{d}_i\|_2^2 = 1.$$

It can be found that Eq. 3.47 shares a lot of similarity to Eq. 3.2 with $\mathbf{X} \leftrightarrow \begin{bmatrix} \mathbf{Y} \\ \sqrt{\rho}\mathbf{X} \end{bmatrix}$ and $\mathbf{D} \leftrightarrow \begin{bmatrix} \mathbf{W} \\ \sqrt{\rho}\mathbf{D} \end{bmatrix}$, except the constraint $\|\mathbf{d}_i\|_2 = 1$. Accordingly we can use the K-SVD algorithm

Algorithm 9 The D-KSVD algorithm

Input : training data $\mathbf{X}$, label of training data $\mathbf{Y}$ and parameter ρ and ϵ
Output : dictionary $\mathbf{D}$, classifier $\mathbf{W}$ and sparse coefficient α

1: Initialize $\mathbf{D}$ and α with the K-SVD algorithm
2: Initialize $\mathbf{W}$ as $(\alpha\alpha^T)^{-1}\alpha\mathbf{Y}$
3: **while not** converged **do**
4:     **for** each dictionary atom $\mathbf{d}_i$ **do**
5:        Use Step 7 to Step 11 in the K-SVD algorithm to update $\mathbf{d}_i$, $\mathbf{w}_i$ and α^i
6:     **end for**
7:     **for** each dictionary atom $\mathbf{d}_i$ **do**
8:        Normalize the $\mathbf{d}_i$, $\mathbf{w}_i$ and α^i using Eq. 3.48
9:     **end for**
10:    Check stop criterion
11: **end while**

---

to solve Eq. 3.47. However, we need to normalize $\mathbf{d}_i$ to satisfy the constraint $\|\mathbf{d}_i\|_2 = 1$:

$$
\begin{aligned}
\mathbf{d}_i &= \frac{\mathbf{d}_i}{\|\mathbf{d}_i\|_2}\forall i \\
\mathbf{w}_i &= \frac{\mathbf{w}_i}{\|\mathbf{d}_i\|_2}\forall i \\
\alpha^i &= \alpha^i \|\mathbf{d}_i\|_2 \forall i.
\end{aligned}
\tag{3.48}
$$

The complete D-KSVD algorithm is summarized in Algorithm 9.

In the label consistent dictionary learning algorithm (LC-KSVD) proposed in Jiang et al. [2011], each dictionary atom is further enforced to represent a subset of the training signals, ideally from a single class, via a label consistence regularization term. Given the sparse coefficient $\alpha \in \mathbb{R}^{K \times N}$, the label consistent regularization term is written as:

$$
\|\mathbf{Q} - \mathbf{A}\alpha\|_F^2,
\tag{3.49}
$$

where $\mathbf{Q} \in \mathbb{R}^{k \times N}$ is the "discriminative" sparse codes of the input for classification and $\mathbf{A}$ is a linear projection matrix. $q_i^j = 1$, if and only if the sparse coefficient (input data) α_i and the dictionary atom $\mathbf{d}_j$ share the same class label; otherwise $q_i^j = 0$. An example of $\mathbf{Q}$ is shown in Eq. 3.50, where we assume $\mathbf{D} = [\mathbf{d}_1, \cdots, \mathbf{d}_6]$, $\alpha = [\alpha_1, \cdots, \alpha_6]$, $\mathbf{d}_1, \mathbf{d}_2, \alpha_1$ and α_2 are from Class

---

Algorithm 10 A classification algorithm with the learned dictionary

---

Input : testing data $\mathbf{x}$, dictionary $\mathbf{D}$ and classifier $\mathbf{W}$
Output : label l

1: Compute the sparse coefficient α
2: Find the label with classifier $l = \max_i (\mathbf{W}\alpha)_i$

1, $\mathbf{d}_3$, $\mathbf{d}_4$, α_3 and α_4 are from Class 2, $\mathbf{d}_5$, $\mathbf{d}_6$, α_5 and α_6 are from the same class.

$$\mathbf{Q} = \begin{bmatrix} 1 & 1 & 0 & 0 & 0 & 0 \\ 1 & 1 & 0 & 0 & 0 & 0 \\ 0 & 0 & 1 & 1 & 0 & 0 \\ 0 & 0 & 1 & 1 & 0 & 0 \\ 0 & 0 & 0 & 0 & 1 & 1 \\ 0 & 0 & 0 & 0 & 1 & 1 \end{bmatrix} \tag{3.50}$$

With the label consistence term, the objective function of LC-KSVD is written as:

$$\mathbf{D}, \alpha, \mathbf{W}, \mathbf{A} \quad : \quad \min_{\mathbf{D}, \alpha, \mathbf{W}, \mathbf{A}} \|\mathbf{Y} - \mathbf{W}^T \alpha\|_F^2 + \rho\|\mathbf{X} - \mathbf{D}^T \alpha\|_F^2 + \lambda\|\mathbf{Q} - \mathbf{A}^T \alpha\|_F^2 \tag{3.51}$$
$$\text{s.t.} \quad \|\alpha_i\|_0 \leq \epsilon \, , \, \|\mathbf{d}_i\|_2^2 = 1.$$

Similar to Eq. 3.47, Eq. 3.51 can be solved via the K-SVD algorithm as:

$$\mathbf{D}, \alpha, \mathbf{W}, \mathbf{A} \quad : \quad \min_{\mathbf{D}, \alpha, \mathbf{W}, \mathbf{A}} \left\| \begin{bmatrix} \mathbf{Y} \\ \sqrt{\rho}\mathbf{X} \\ \sqrt{\lambda}\mathbf{Q} \end{bmatrix} - \begin{bmatrix} \mathbf{W} \\ \sqrt{\rho}\mathbf{D} \\ \sqrt{\lambda}\mathbf{A} \end{bmatrix} \alpha \right\|_F^2 \tag{3.52}$$
$$\text{s.t.} \quad \|\alpha_i\|_0 \leq \epsilon \, , \, \|\mathbf{d}_i\|_2^2 = 1.$$

After the dictionary is learned, the normalization process in Eq. 3.48 is also applied to make sure each dictionary atom has unit ℓ_2 length.

After the dictionary $\mathbf{D}$ and classifier $\mathbf{W}$ are learned from the discriminative dictionary learning algorithms, we can apply them in classification tasks. Algorithm 10 illustrates a typically way of doing this.

Besides class labels of the training data, pairwise relationship was studied in Guo et al. [2013] for discriminative dictionary learning, where the sets of the same and different pairs, $\mathcal{S}$ and $\mathcal{D}$ respectively, are given, from which a matrix $\mathbf{M}$ is defined to encode such information:

$$\mathbf{M}_{i,j} = \begin{cases} +1 & , \text{if } (i, j) \in \mathcal{S} \\ -1 & , \text{if } (i, j) \in \mathcal{D} \\ 0 & \text{otherwise.} \end{cases} \tag{3.53}$$

Accordingly, the objective function is written as:

$$\mathbf{D}, \mathbf{W}, \alpha \quad : \quad \min_{\mathbf{D}, \mathbf{W}, \alpha} \|\mathbf{X} - \mathbf{D}\alpha\|_F^2 + \gamma \|\alpha\|_1 + \mu \|\mathbf{Y} - \mathbf{W}\alpha\|_F^2 + \lambda \|\mathbf{W}\|_F^2 \qquad (3.54)$$
$$+ \beta \sum_{i,j} \mathbf{M}_{i,j} \|\alpha_i - \alpha_j\|_2^2$$

$$: \quad \min_{\mathbf{D}, \mathbf{W}, \alpha} \|\mathbf{X} - \mathbf{D}\alpha\|_F^2 + \gamma \|\alpha\|_1 + \mu \|\mathbf{Y} - \mathbf{W}\alpha\|_F^2 + \lambda \|\mathbf{W}\|_F^2 \qquad (3.55)$$
$$+ \beta \operatorname{tr}(\alpha^T \mathbf{L} \alpha),$$

where $\mathbf{L}$ is the Laplacian matrix for $\mathbf{M}$. A block coordinate descent algorithm was used for solving the problem.

Besides linear classification and squared loss [Jiang et al., 2011, Pham and Venkatesh, 2008, Zhang and Li, 2010a], other types of classifiers and loss functions have also been studied. For example, in Mairal et al. [2009c] a bilinear classifier and the logistic loss were used for the discriminative term $f(\mathbf{x}, \alpha, \theta) = \mathbf{x}^T \mathbf{W}\alpha + b$, where $\theta = \{\mathbf{W} \in \mathbb{R}^{d \times k}, b \in \mathbb{R}\}$ is the classifier parameter. The logistic loss function, which is defined as $\mathcal{C}(x) = \log 1 + e^{-x}$, is more robust for classification while being differentiable. It was also shown in Boureau et al. [2010] through experiment that, learning a supervised discriminative dictionary for sparse coding is able to improve the performance of the bag-of-words model in object recognition, where the linear classifier and logistic loss were used for analysis.

The classifier $\mathbf{W}$ can be interpreted as a reflection of the importance of each dictionary atom to different classes. In Yang et al. [2013a], this interpretation is especially clear, since $\mathbf{W}$ is required to be nonnegative while the sum of $\mathbf{W}$ for each class is 1. Let $\mathbf{X}_i$ be the training data of Class i, the dictionary is learned via solving the following problem:

$$\mathbf{D}, \mathbf{W}, \alpha \quad : \quad \min_{\mathbf{D}, \mathbf{W}, \alpha} \sum_i \|\mathbf{X}_i - \mathbf{D}\mathcal{D}(\mathbf{w}_i)\alpha_i\|_F^2 + \lambda \|\alpha_i\|_1 + \gamma \|\alpha_i - \frac{1}{N_i} \sum_{j:l_j=l_i} \alpha_j\|_F^2$$
$$+ \mu \sum_{j \neq i} \sum_p \sum_{q \neq p} w_i^p (\mathbf{d}_p^T \mathbf{d}_q)^2 w_j^q \qquad (3.56)$$
$$\text{s.t.} \quad w_i^p \geq 0, \ \sum_p w_i^p = 1,$$

where $\mathcal{D}(\cdot)$ is the diagonal operator. $\|\alpha_i - \frac{1}{N_i} \sum_{j:l_j=l_i} \alpha_j\|_F^2$ forces the coefficients of data of the same class to be similar and $\sum_i \sum_{j \neq i} \sum_p \sum_{q \neq p} w_i^p (\mathbf{d}_p^T \mathbf{d}_q)^2 w_j^q$ requires that, if two dictionary atoms are similar (i.e., $(\mathbf{d}_p^T \mathbf{d}_q)^2$ or the coherence is large), then they should have different importance to the classes (i.e., $w_i^p w_j^q$ is small), with w_i^p being the p−th element of the vector $\mathbf{w}_i$.

For classification, two schemes were proposed, given the learned dictionary.

Local classifier: the label of a testing data point $\mathbf{x}$ is found by solving the following problem $i : \min_i \min_\alpha \|\mathbf{x}\mathbf{D}\mathcal{D}(\mathbf{w}_i)\alpha\|_2^2 + \lambda \|\alpha\|_1$;

Global classifier: when training data of each class are limited, sparse coefficient is first computed by solving the problem $\min_\alpha \|\mathbf{x} - \mathbf{DD}(\frac{1}{C}\sum_i \mathbf{w}_i)\alpha\|_2^2 + \lambda\|\alpha\|_1$ and then the label is decided by minimizing $i : \min_i \|\mathbf{x} - \mathbf{DD}(\mathbf{w}_i)\alpha\|_2^2$.

Clearly, the various dictionary learning algorithms presented previously may behave differently, when applied to the same dataset. Figure 3.3 shows examples of the sparse coefficients obtained with some of the algorithms, which may help a reader gain a little more insight into the potential differences of the sparse representations under the dictionaries.

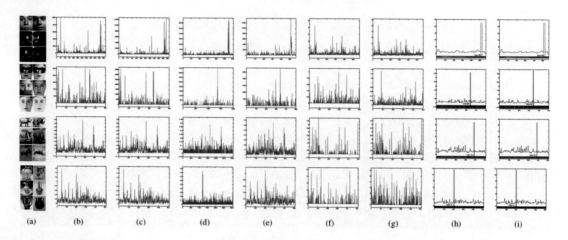

Figure 3.3: Examples of sparse codes using different dictionary learning methods on the three evaluated datasets. X-axis indicates the dimensions of sparse codes, Y-axis indicates a sum of absolute sparse codes for different testing samples from the same class. The curves in 1st, 2nd, 3rd, and 4th row correspond to class 35 in Extended YaleB (32 testing frames), class 69 in AR Face (6 testing frames), class 78 (29 testing frames) and class 41 (55 testing frames) in Caltech 101, respectively. (a) Sample images from these classes. The sparse coding approaches include: (b) K-SVD [Aharon et al., 2005]; (c) D-KSVD [Zhang and Li, 2010a]; (d) SRC [Wright et al., 2009c]; (e) SRC* [Wright et al., 2009c]; (f) LLC [Wang et al., 2010] (30 local bases); (g) LLC [Wang et al., 2010] (70 local bases); (h) LC-KSVD [Jiang et al., 2011] (without classification term); (i) LC-KSVD [Jiang et al., 2011]. Each color from the color bars in (h) and (i) represents one class for a subset of dictionary items. The black dashed lines demonstrate that the curves are highly peaked in one class. The two examples from Caltech 101 demonstrate that a large intra-class difference is enforced between classes via LC-KSVD. The figure is best viewed in color and 600% zoom in. This figure is from Jiang et al. [2011].

3.2.2   IMPLICIT DISCRIMINATIVE DICTIONARY LEARNING

In the previous subsection, we presented several explicit discriminative dictionary learning algorithms, where the classifier $\mathbf{W}$ and the dictionary $\mathbf{D}$ are learned jointly to achieve the discriminative capability. In contrast, in implicit discriminative dictionary learning, the goal is to embed

---

Algorithm 11 The classification algorithm with the learned dictionary

---

Input : testing data $\mathbf{x}$, $\mathbf{D}$
Output : label l

1: **for** each subdictionary $\mathbf{D}_i$ **do**
2:   Compute the sparse coefficient α
3:   Compute the reconstruction error $r_i = \|\mathbf{x} - \mathbf{D}_i\alpha\|_2$
4: **end for**
5: Assign the label as l : $\min_l r_l$

---

some discriminative capability into a learned dictionary, but no classifier is explicitly learned at the same time.

In Mairal et al. [2008a], a set of sub-dictionaries $\mathbf{D} = \{\mathbf{D}_1, \mathbf{D}_2, \cdots, \mathbf{D}_N\}$ is learned, where each sub-dictionary is associated to a class S_i. To measure the quality of the association, the reconstruction error is used: signals of class S_i should be represented well by dictionaries $\mathbf{D}_i$ with a small reconstruction error whereas large reconstruction errors may incur with all the other subdictionaries $\mathbf{D}_j$ $j \neq i$. This problem can be formulated as:

$$\{\mathbf{D}_i\}_i^N : \min_{\{\mathbf{D}_i\}_i^N} \sum_j C_{y_j}^\lambda (\{R^*(\mathbf{x}_j, \mathbf{D}_i)\}_i^N) + \gamma R^*(\mathbf{x}_j, \mathbf{D}_{y_j}), \qquad (3.57)$$

where $C_i^\lambda(\{z_i\}_i^N) = \log \sum_j^N e^{-\lambda(y_j - y_i)}$ is the softmax discriminative cost function and $R^*(\mathbf{x}, \mathbf{D}) = \min_{\alpha:\|\alpha\|_0 \leq T} \frac{1}{2}\|\mathbf{x} - \mathbf{D}\alpha\|_2$ is the minimal reconstruction error for signal $\mathbf{x}$ with dictionary $\mathbf{D}$.

From Eq. 3.57, we can find that not only each of the learned dictionaries $\mathbf{D}_i$ is capable of sparse-reconstructing the signals of class S_i with a small residual but also this residual is smaller than the reconstruction with all the other $\mathbf{D}_j$, $j \neq i$. Accordingly, the learned dictionaries can be applied in classification, without explicit learning of the classifier as in the D-KSVD algorithm. The classification algorithm is shown in Algorithm. 11.

Equation 3.57 is non-convex and non-differential. To alleviate this problem, the soft-max function is first approximated piece-wise linearly and then the problem of updating $\mathbf{D}_i$ is formulated as:

$$\mathbf{D}_i : \min_{\mathbf{D}_i} \sum_l \sum_{k:y_k=l} w_k R(\mathbf{x}_k, \mathbf{D}_i, \alpha_k^i) \qquad (3.58)$$

with

$$w_k = \frac{\partial C_{y_k}^\lambda}{\partial i}(\{R^*(\mathbf{x}_j, \mathbf{D}_i)\}_i^N) + \gamma \delta(i, y_k). \qquad (3.59)$$

Algorithm 12 The algorithm proposed in Mairal et al. [2008a]

Input : training data $\mathbf{X}$, label of training data $\mathbf{Y}$ and parameter ρ and ϵ
Output : dictionary $\mathbf{D}$, classifier $\mathbf{Y}$ and sparse coefficient α

 1: Initialize $\mathbf{D}$ and α
 2: **while not** converged **do**
 3:     **for** each subd-ictionary $\mathbf{D}_i$ **do**
 4:        **for** each dictionary atom $\mathbf{d}_{i,j}$ **do**
 5:           Find the set of signals which utilize $\mathbf{d}_{i,j}$, i.e., $\Omega = \{k \,|\, \alpha_k^{i,j} \neq 0\}$
 6:           For each signal $\mathbf{x}_k$ in Ω, compute the residual $\mathbf{r}_k^i = \mathbf{x}_k - \mathbf{D}_i \alpha_k^i$
 7:           Compute weight vector w_l with Eq. 3.59
 8:           Update the atom $\mathbf{d}_{i,j}$ and coefficient $\alpha^{i,j}$ by solving Eq. 3.60
 9:        **end for**
10:     **end for**
11:     Check stop criterion
12: **end while**

Similar to the idea of K-SVD, the atom $\mathbf{d}_{i,j}$ and coefficient $\alpha^{i,j}$ can be updated via solving the following problem:

$$\mathbf{d}_{i,j}, \alpha^{i,j} \quad : \quad \min_{\mathbf{d}_{i,j}, \alpha^{i,j}} \sum_{k \in \Omega} w_k \|\mathbf{r}_k^i - \mathbf{d}_{i,j} \alpha_k^{i,j}\|_2^2 \qquad (3.60)$$

$$\text{s.t.} \quad \|\mathbf{d}_{i,j}\|_2 = 1. \qquad (3.61)$$

The final algorithm is presented in Algorithm 12.

A similar idea was also found in Kong et al. [2013], where a set of dictionaries is learned from data of multiple classes, such that the reconstruction error of the sparse representation with the dictionary of the corresponding class, namely intra-class reconstruction error, is minimized, while the reconstruction error of the sparse representation with the dictionary of other classes, namely inter-class reconstruction error is maximized. A K-SVD-like algorithm was proposed to learn the dictionaries under such constraints. A general formulation and optimization framework was proposed in Mairal et al. [2012] and Zhang et al. [2013b] for supervised dictionary learning, where a wide set of convex loss functions can be used according to the desired use of the dictionary. For example, squared error is used for regression, while logistic loss is used for classification..

Instead of relying on the reconstruction error as in Mairal et al. [2008a], inter-class dictionary coherence is used for discriminative dictionary learning in Ramirez et al. [2010]. Given the training data $\{\mathbf{X}_i\}$ from different classes $i = 1, \cdots, K$, a set of dictionaries is learned, with each dictionary $\mathbf{D}_i$ optimized for representing data $\mathbf{X}_i$ of Class i:

$$\{\mathbf{D}_i, \alpha_i\} : \min_{\{\mathbf{D}_i, \alpha_i\}} \sum_i \|\mathbf{X}_i - \mathbf{D}_i \alpha_i\|_F^2 + \lambda \|\alpha_i\|_1 + \eta \sum_{i \neq j} \|\mathbf{D}_i^T \mathbf{D}_j\|_F^2, \qquad (3.62)$$

where $\sum_{i \neq j} \|\mathbf{D}_i^T \mathbf{D}_j\|_F^2$ enforces the coherence of the dictionaries of different classes to be small. Given a testing data $\mathbf{x}$, its label can be found by solving the following problem:

$$i : \min_i \min_\alpha \|\mathbf{x} - \mathbf{D}_i \alpha\|_2^2 + \lambda \|\alpha\|_1. \qquad (3.63)$$

However, it was found in Ramirez et al. [2010] that, the dictionaries $\mathbf{D}_i$ share a lot of common atoms that are used often and their corresponding coefficients have large absolute values. This makes the reconstruction errors with different dictionaries similar. Those atoms can be detected via checking $|\mathbf{D}_i \mathbf{D}_j|$, which typically have large values (0.95 was used in Ramirez et al. [2010] as the threshold to identify such shared atoms). By removing the shared atoms for reconstruction, the discriminative performance can be improved. The method was also extended for clustering problems, where the dictionary can be initialized with spectral clustering.

Intra-class dictionary coherence was also used in Chi et al. [2013] for dictionary learning, where the signal can be represented by a small sets of blocks of dictionary atoms and the intra-block coherence of the dictionary is minimized. The dictionary consists of multiple sub-dictionaries (or blocks), where sub-dictionary $\mathbf{D}_i$ is optimized for sparse-representing data of Class i with minimal reconstruction error but not for data of other classes. The objective function is written as:

$$\mathbf{D}, \alpha : \min_{\mathbf{D}, \alpha} \frac{1}{2} \sum_i \|\mathbf{X}_i - \mathbf{D}_i \alpha_i\|_F^2 + \lambda \|\alpha_i\|_{1,p} + \gamma \|\mathbf{D}\|_F^2 + \beta \sum_i \sum_{\mathbf{d}_p, \mathbf{d}_q \in \mathbf{D}_i} \|\mathbf{d}_p^T \mathbf{d}_q\|_2^2, \qquad (3.64)$$

where in $\|\alpha_i\|_{1,p}$, the group sparsity constraint is applied to the sparse coefficient to enforce that the same set of dictionary atoms are shared for data of one class and $\sum_{\mathbf{d}_p, \mathbf{d}_q \in \mathbf{D}_i} \|\mathbf{d}_p^T \mathbf{d}_q\|_2^2$ measures the intra-block dictionary coherence for Block $\mathbf{D}_i$. In Chi et al. [2013], $(1, 1)$ and $(1, 2)$ were proposed for the group sparsity constraint on the sparse coefficient. In particular, with $(1, 1)$ the group sparsity constraint is separable. However, one limitation of the method is that, the blocks/groups need to be manually specified. Similar idea was also used in Luo et al. [2013] for action recognition from 3D skeleton trajectories.

Mutual information (or entropy) was utilized in discriminative dictionary learning in Qiu et al. [2011, 2014]. The dictionary is built by greedily selecting atoms from an initial big dictionary (from K-SVD), where in each round, the selected atom maximizes the reduction of entropy about the remaining dictionary and the reduction of the entropy of the labels. The relationship between dictionary $\mathbf{D}$ and training data $\mathbf{X}$ was model via Gaussian process. The conditional probability of the sparse coefficient $\alpha_\mathbf{d}$ for atom $\mathbf{d}$ given the dictionary $\mathbf{D}$ is a Gaussian distribution with closed-form variance:

$$\mathbb{V}(\mathbf{d}|\mathbf{D}) = \mathcal{K}(\mathbf{d}, \mathbf{d}) - \mathcal{K}^T(\mathbf{d}, \mathbf{D}) \mathcal{K}^{-1}(\mathbf{D}, \mathbf{D}) \mathcal{K}(\mathbf{d}, \mathbf{D}), \qquad (3.65)$$

where $\mathcal{K}(\mathbf{d}, \mathbf{D})$ is the variance between $\mathbf{d}$ and $\mathbf{D}$. Accordingly, the entropy of $\mathbf{d}$ given $\mathbf{D}$ can be computed as:

$$H(\mathbf{d}|\mathbf{D}) = \frac{1}{2} \log(2\pi e \mathbb{V}(\mathbf{d}|\mathbf{D})). \tag{3.66}$$

The entropy of the labels given the dictionary atoms can be computed similarly from labels of the training data. The sparse coefficient represents the training data and thus can be associated with the labels. Then the entropy of the labels given the dictionary atom can be computed from the distribution of the labels and the sparse coefficient of each dictionary atom.

The Fisher criterion, which measures the ratio of inter-class scatter $S_b()$ and intra-class scatter $S_w()$, was used for discrminative dictionary learning in Yang et al. [2011a]. The Fisher criterion is applied for the sparse coefficient α:

$$f(\alpha) = \operatorname{tr}(S_w(\alpha)) - \operatorname{tr}(S_b(\alpha)) + \eta\|\alpha\|_F^2. \tag{3.67}$$

By introducing the Fisher criterion term, the objective function of Yang et al. [2011a] can be written as:

$$\mathbf{D}, \alpha : \min_{\mathbf{D},\alpha} \sum_i \|\mathbf{X}_i - \mathbf{D}\alpha_i\|_F^2 + \|\mathbf{X}_i - \mathbf{D}_i\alpha_i^i\|_F^2 + \sum_{j\neq i} \|\mathbf{D}_j\alpha_i^j\|_F^2 + \lambda f(\alpha) + \gamma\|\alpha\|_1, \tag{3.68}$$

where $\mathbf{D} = \{\mathbf{D}_i\}$ is the dictionary and α_i^j refers to the coefficient corresponding to $\mathbf{D}_i$. Equation 3.68 enforces (1) the dictionary represents the data with small reconstruction error; (2) each subdictionary represents the data of the corresponding class with small reconstruction error; (3) the subdictionary receives small sparse coefficients for data of different classes and also (4) the sparse coefficient maximizes the Fisher criterion. Two examples of dictionaries learned from images of Digit 8 and 9 from the USPS dataset can be found in Fig. 4 of Yang et al. [2011a]. After the dictionary is learned, it can be used in classification. Two classification schemes were proposed in Yang et al. [2011a], similar to what was done in Yang et al. [2013a].

For most of the algorithms presented above, the label information for the training data is assumed to be noise-free. In practice, this may be invalid. In Chen et al. [2013a], an algorithm for learning dictionaries from data with ambiguous labels was proposed. Similar to most of the algorithms described earlier, the dictionary is assumed to consist of multiple sub-dictionaries where each of them corresponds to data of one class. The algorithm has two steps. The first step computes the label confidence for each data $\mathbf{x}_i$ vs each label j according to the reconstruction error with each of the sub-dictionaries:

$$P_i^j = \frac{\beta_j e^{-\frac{e_i^j}{2\sigma_j}}}{\sum_{k\in L_i} \beta_k e^{-\frac{e_i^k}{2\sigma_k}}}, \tag{3.69}$$

where L_i is the ambiguous label set for data $\mathbf{x}_i$, β_j is the frequency of Label j in the training set and σ_j is the mean of the reconstruction error for Label j. e_i^j is computed as:

$$e_i^j = \|\mathbf{x}_i - \mathbf{D}_j(\mathbf{D}_j^T\mathbf{D}_j)^{-1}\mathbf{D}_j^T\mathbf{x}_i\|_2^2 \tag{3.70}$$

which can be viewed as the reconstruction error when x_i is projected onto the subspace spanned by D_j. In the second step, the sub-dictionaries are updated with data of each class. Considering the ambiguity of the labels, two types of schemes for assigning labels to the data according to the label confidence were proposed. For hard-decision, the data is assigned to the label, whose confidence is the highest. For soft-decision, each training data point is assigned to all labels softly, whose weight is the label confidence. Thus, the higher label confidence it is, the more important is the training data to that class.

Most of the existing discriminative dictionary learning algorithms also assume that data of different classes are independently distributed, which might not be true for a real application. Modeling the dependencies of different classes could facilitate classification, as shown in Siyahjani and Doretto [2013], where the co-occurrences of different classes are computed to encode such dependencies, which are then utilized in context-aware dictionary learning. It was shown that the learned dictionary, which is encoded with co-occurrence of different classes in the image, can facilitate the object detection task. This context-aware dictionary learning idea was formulated by using an objective function that consists of three terms: the reconstruction error term, the classification error term, and the context-aware term.

In Chiang et al. [2013], a dictionary learning algorithm for data with multiple attributes was proposed. To utilize multiple attributes, a distance metric was proposed to incorporate both data and attribute similarities. The objective function constrains the dictionary to be both compact, reconstructive and label consistent.

3.3   JOINT LEARNING OF MULTIPLE DICTIONARIES

In the previous two sections, we discussed the problem of learning a single dictionary, either for reconstruction or for classification, and presented several algorithms on this regard. In some applications such as image super-resolution and transfer learning, where two domains are typically considered at the same time, more than one dictionary may need to be learned simultaneously. In this section, we discuss the problem of learning multiple dictionaries simultaneously (which will be referred to as joint dictionary learning) and present some of the most important algorithms on this problem.

3.3.1   LEARNING DICTIONARIES FROM MULTIPLE CLUSTERS

In algorithms like K-SVD, a single dictionary is trained to represent all of the training data. If the training data contain some clustering structure, individual dictionaries may be trained to cover the underlying clusters, which could improve the effectiveness in representation of the data. This intuition has led to some multi-dictionary learning algorithms. In Li and Fang [2010], the input data of image patches are divided into multiple groups according to their appearances and one dictionary is learned from each group of patches. By utilizing the learned set of dictionaries, each of which is optimal for some specific type of patches, improved image denoising performance was observed. A similar idea was found in Feng et al. [2011b], where the K-subspace algorithm was used for clustering the data into K groups. In the K-subspace algorithm, each data sample $\mathbf{x}$ is assigned to the closest subspace where its distances to subspace $\mathcal{S}_k$ is computed as $\|\mathbf{x} - \mathbf{U}_k \mathbf{U}_k^T \mathbf{x}\|_2$, with $\mathbf{U}_k$ the basis of subspace $\mathcal{S}_k$. Then for each subspace a dictionary is learned from its samples via the K-SVD algorithm. In addition, the number of subspaces K can be decided adaptively from the data by initializing $k = 1$ and increasing k by 1 until the subspace has become too small.

A clustering framework was proposed in Sprechmann and Sapiro [2010] based on sparse modeling and dictionary learning. A data sample is no longer represented by its closest cluster center, as in K-means, but instead is sparsely represented by atoms of the corresponding sub-dictionary. Thus, the training data is modeled as the union of the learned low-dimensional subspaces. Clustering is formulated as an optimization problem with sparsity constraint and an optimization algorithm was presented. The algorithm was evaluated on digit classification and texture segmentation, with promising results reported. In Chen et al. [2013b], this problem was considered for clustering image, where the dictionaries and clustering are learned in the radon transform domain to achieve shift-rotation-scale invariant, which is important for content-based image retrieval (CBIR).

A similar problem was also studied in Ma et al. [2012], where each of the dictionaries is required to be low rank and the correlation between a dictionary and the data of other classes is required to be small. Formally, this problem is formulated as:

$$\mathbf{D}, \mathbf{E}, \alpha \quad : \quad \min_{\mathbf{D}, \mathbf{E}, \alpha} \sum_i \|\alpha_i\|_1 + \gamma \|\mathbf{D}_i\|_* + \beta \|\mathbf{E}_i\|_1 + \lambda r(\mathbf{D}_i) \tag{3.71}$$
$$\text{s.t.} \quad \mathbf{X}_i = \mathbf{D}\alpha_i + \mathbf{E}_i, \mathbf{X}_i = \mathbf{D}_i \alpha_i^i + \mathbf{E}_i,$$

where $\mathbf{X}_i$ is the data of Group i, $\mathbf{E}_i$ is the sparse corruptions or noise and $\|\cdot\|_*$ is the nuclear norm for measuring the rank of a matrix. The constraint $\mathbf{X}_i = \mathbf{D}_i \alpha_i^i + \mathbf{E}_i$ ensures that the dictionary $\mathbf{D}_i$ is able to reconstruct the data of Class i with minimum error. In order to minimize $r(\mathbf{D}_i) = \sum_{j \neq i} \|\mathbf{D}_i \alpha_j^i\|_F^2$, α_j^i should be close to zero for $j \neq i$.

This problem was studied in Abolghasemi et al. [2012] for blind source separation (BSS). In BSS, one aims to find out the source $\mathbf{X}$ and the mixing matrix $\mathbf{A}$ from the mixed observation $\mathbf{Y}$, subject to some noise $\mathbf{V}$, i.e., $\mathbf{Y} = \mathbf{A}\mathbf{X} + \mathbf{V}$. BSS was formulated into the problem of learning multiple dictionaries, where each dictionary represents one source. Note that the mixing matrix $\mathbf{A}$ (or grouping information) is unknown. This is a key difference between this problem and those in Feng et al. [2011b], Li and Fang [2010], and Ma et al. [2012], where the grouping information is assumed known (either given or from some clustering algorithms) and fixed for dictionary learning. Considering this, a method was proposed in Abolghasemi et al. [2012] to learn the mixing matrix and the dictionaries jointly, where the mixing matrix is updated according to the learned dictionary and the dictionary is updated from the newly-estimated mixing matrix. Considering a set of $2D$ images, the problem is then formulated as follows:

$$\mathbf{D}, \mathbf{A}, \alpha, \mathbf{X} \min_{\mathbf{D}, \mathbf{A}, \alpha, \mathbf{X}} \lambda \|\text{vec}(\mathbf{Y}) - (\mathbf{I} \otimes \mathbf{A})\text{vec}(\mathbf{X})\|_F^2 + \sum_i \|\mathcal{R}_i(\text{vec}(\mathbf{X})) - \mathbf{D}_i \alpha\|_F^2 + \mu_i \|\alpha_i\|_0,$$

$$\tag{3.72}$$

where $\mathbf{Y} \in \mathbb{R}^{n \times N}$ is the set of vectorized images, vec($\cdot$) vectorizes a matrix, $\otimes$ is the Kronecker product, and $\mathcal{R}_i(\cdot)$ extracts patches from the i_{th} source. In this problems, the patches of each source (i.e., an image) can be sparsely represented by one dictionary. A block coordinate descent method was proposed for solving the above problem. For $\mathbf{X}$, it can be computed as vec$(\mathbf{X}) = (\lambda(\mathbf{I} \otimes \mathbf{A})^T(\mathbf{I} \otimes \mathbf{A}) + \sum_i \mathcal{R}_i^T \mathcal{R}_i)^{-1}(\lambda(\mathbf{I} \otimes \mathbf{A})^T \text{vec}(\mathbf{Y}) + \sum_i \mathcal{R}_i^T \mathbf{D}_i \alpha_i)$, the mixing matrix is updated as $\mathbf{A} = \mathbf{Y}\mathbf{X}^T(\mathbf{X}\mathbf{X}^T)^{-1}$.

3.3.2   LEARNING DICTIONARIES FROM MULTIPLE SUBSPACES

In the previous subsection, we discussed the algorithms that learn multiple dictionaries from data forming multiple groups such that each dictionary is optimized for a specific group. In this subsection, we introduce another type of joint dictionary learning algorithms, in which the learned dictionaries represent the data of multiple groups in a different way. In examples like signal processing, when processing signals of an ensemble, the signals may be found to share some common components, besides their signal-specific parts. Several sparse representation algorithms have been proposed to extract the common components and signal-specific components from the signals of an ensemble. A good example of this sort is distributed compressive sensing of multiple signals [Baron et al., 2009]. Many benefits can be drawn from such a decomposition of an ensemble of signals, such as obtaining better compression rate [Baron et al., 2009] and being able to extract more relevant features [Nagesh and Li, 2009].

In distributed compressive sensing (DCS or joint sparsity models, JSM), each signal $\mathbf{x}_i$ of an ensemble can be represented by the sum of the common component $\mathbf{z}_c$ and the innovation component $\mathbf{z}_i$, i.e.,

$$\mathbf{x}_i = \mathbf{z}_c + \mathbf{z}_i. \tag{3.73}$$

The common component is shared by all signals and the innovation component $\mathbf{z}_i$ is specific to signal $\mathbf{x}_i$. With the sparsity assumption, the common and innovation components can be sparsely represented under some dictionaries:

$$\mathbf{z}_c = \mathbf{D}_c \alpha_c \tag{3.74}$$
$$\mathbf{z}_i = \mathbf{D}_i \alpha_c, \tag{3.75}$$

where $\mathbf{D}_c$ and $\mathbf{D}_i$ are the dictionaries for the common component and the innovation component respectively and α_c and α_i are the corresponding sparse coefficients. For example, the DCT matrix was studied in Nagesh and Li [2009]. Accordingly, the joint sparse model can be represented as:

$$\mathbf{x}_i = \mathbf{D}_c \alpha_c + \mathbf{D}_i \alpha_i \ \forall i \tag{3.76}$$

$$\begin{bmatrix} \mathbf{x}_1 \\ \mathbf{x}_2 \\ \cdots \\ \mathbf{x}_n \end{bmatrix} = \begin{bmatrix} \mathbf{D}_c & \mathbf{D}_1 & 0 & \cdots & 0 \\ \mathbf{D}_c & 0 & \mathbf{D}_2 & \cdots & 0 \\ \cdots & \cdots & \cdots & \cdots \\ \mathbf{D}_c & 0 & 0 & \cdots & \mathbf{D}_n \end{bmatrix} \begin{bmatrix} \alpha_1 \\ \alpha_2 \\ \cdots \\ \alpha_n \end{bmatrix}. \tag{3.77}$$

Different constraints on α_c and α_i have been discussed in Baron et al. [2009], including: **JSM-1** (where only the common component is required to be sparse), **JSM-2** (where only the innovation components are sparse and in addition their support are required to be the same), and **JSM-3** (where only the innovation components are required to be sparse).

The joint sparsity models have been used in Nagesh and Li [2009] for face recognition, where the common component captures the neutral face image of each subject, and the innovation components capture the expression variation in the face images. The common and innovation components are sparsely represented under the DCT matrix. It was shown that a new test image of a subject can be fairly well approximated using only the two components from the same subject. Based on this, an efficient expression invariant classifier was proposed and achieved good performance.

Although several distributed sparse coding algorithms were discussed in Baron et al. [2009], learning the dictionary $\mathbf{D}_c$ and $\mathbf{D}_i$ was not part of the task. In Zhang and Li [2010b, 2012], an algorithm was proposed to learn the common, condition and residual components for sparsely representing images of multiple classes. Mathematically it can be written as:

$$\mathbf{X}_{i,j} = \mathbf{C}_i + \mathbf{A}_{i,j} + \mathbf{E}_i^j \ \forall (i,j), \tag{3.78}$$

where $\mathbf{X}_{i,j} \in \mathbb{R}^{h \times w}$ is the j_{th} instance of Class i, $\mathbb{C} = \{\mathbf{C}_i\}$ are the common components, $\mathbb{A} = \{\mathbf{A}_{i,j}\}$ are the condition components, and $\mathbb{E} = \{\mathbf{E}_{i,j}\}$ are the innovative components. In this formulation, data of each class i share a common component $\mathbf{C}_i$.

To make the problem well conditioned, the condition components $\mathbb{A}$ are required to be low-rank matrices and residual components $\mathbb{E}$ are sparse matrices. It has been shown that the reconstruction of the image with first few singular values and the corresponding singular vectors often capture the global information of the image [Liu et al., 2008], e.g., illumination conditions, structured patterns, which can be represented by a low-rank matrix. Accordingly, for a given set of images of different classes, we can find the following:

$\mathbf{C}_i$: a matrix representing the common information of images for Subject i, the common components (Fig. 3.4 (a));

$\mathbf{A}_{i,j}$: a low-rank matrix capturing the global information of the image $\mathbf{X}_{i,j}$, e.g., illumination conditions (Fig. 3.4 (b))); and

$\mathbf{E}_{i,j}$: a sparse matrix pertaining to image-specific details such as expression conditions or noise with sparse support in the images.

In addition, by assuming that the images are properly organized so that the j-th instance of all classes have similar global variations, i.e., $\mathbf{A}_{i,j} = \mathbf{A}_{k,j} \ \forall \ j$ and $k \neq i$, the problem can be formulated as:

$$\mathbb{C}, \mathbb{A}, \mathbb{E} = \underset{\mathbb{C}, \mathbb{A}, \mathbb{E}}{\operatorname{argmin}} \sum_{i,j} \|\mathbf{A}_j\|_* + \lambda_{i,j} \|\mathbf{E}_{i,j}\|_1 \text{ s.t. } \mathbf{X}_{i,j} = \mathbf{C}_i + \mathbf{A}_j + \mathbf{E}_{i,j}, \ \forall \mathbf{X}_{i,j} \in \mathbb{X}, \tag{3.79}$$

where $\|\mathbf{A}_j\|_* = \sum_i \sigma_i(\mathbf{A}_j)$ is the nuclear norm, $\|\mathbf{E}_{i,j}\|_1 = \sum_{p,q} |\mathbf{E}_{i,j}(p,q)|$ is the ℓ_1 norm and $\mathbb{E} = \{\mathbf{E}_{i,j}\}_{i,j=1}^{N,M}$. Note that, unlike Wright et al. [2009b] in which a set of images are stacked as vectors of a low-rank matrix, the image is not converted to a vector in the decomposition stage. The examples of decomposition on the extended YaleB dataset are shown in Fig. 3.4.

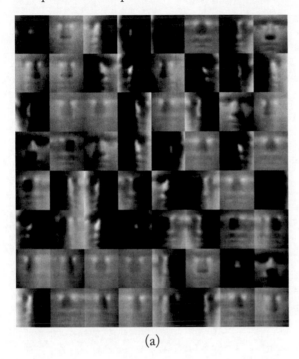 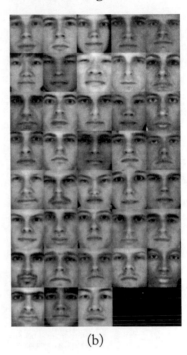

(a)   (b)

Figure 3.4: The decomposition of the extended YaleB dataset. All 2432 images were used, which contain 38 subjects (b) and 64 illumination conditions (a).

To solve the problem in Eq. 3.79, augmented Lagrange multiplier and block coordinate descent were used.With augmented Lagrange multiplier, Eq. 3.79 can be written as:

$$
\mathbb{C}, \mathbb{A}, \mathbb{E} = \underset{\mathbb{C}, \mathbb{A}, \mathbb{E}}{\operatorname{argmin}} \sum_{i,j} \|\mathbf{\Lambda}_j\|_* + \lambda_{i,j} \|\mathbf{E}_{i,j}\|_1 + \frac{\mu_{i,j}}{2} \|\mathbf{X}_{i,j} - \mathbf{C}_i - \mathbf{A}_j - \mathbf{E}_{i,j}\|_F^2
$$
$$
+ \quad \langle \mathbf{Y}_{i,j}, \mathbf{X}_{i,j} - \mathbf{C}_i - \mathbf{A}_j - \mathbf{E}_{i,j} \rangle, \tag{3.80}
$$

where $\mathbf{Y}_{i,j}$ is the Lagrange multiplier, $\lambda_{i,j}$ and $\mu_{i,j}$ are scalars controlling the weight of sparsity and reconstruction error accordingly. A block coordinate descent algorithm was designed for solving the above equation, with each iterative step solving a convex optimization problem [Candes and Plan, 2009, Wright et al., 2009b] for one of the $\mathbb{C}$, $\mathbb{A}$ and $\mathbb{E}$ while fixing others. The algorithm was described in Algorithm 13.

Sub-solution 1: For finding an optimal $E_{i,j}$ in the t-th iteration, the problem can be written as

$$
\mathbf{E}_{i,j} = \underset{\mathbf{E}_{i,j}}{\operatorname{argmin}} \lambda_{i,j} \|\mathbf{E}_{i,j}\|_1 + \frac{\mu_{i,j}}{2} \|\mathbf{X}_{i,j}^E - \mathbf{E}_{i,j}\|_F^2 + \langle \mathbf{Y}_{i,j}, \mathbf{X}_{i,j}^E - \mathbf{E}_{i,j} \rangle \tag{3.81}
$$

with $\mathbf{X}_{i,j}^E = \mathbf{X}_{i,j} - \mathbf{C}_i - \mathbf{A}_j$. According to Hale et al. [2008], it can be solved as:

$$\mathbf{E}_{i,j} = S_{\frac{\lambda}{\mu_{i,j}}}(\mathbf{X}_{i,j}^E + \frac{1}{\mu_{i,j}}\mathbf{Y}_{i,j}), \tag{3.82}$$

where $S_\tau(\mathbf{X}) = sign(\mathbf{X}) \cdot max(0, |\mathbf{X}| - \tau)$.

Sub-solution 2: For finding an optimal $\mathbf{A}_k$ in the t-th iteration, the problem can be written as

$$\mathbf{A}_j = \underset{\mathbf{A}_j}{\operatorname{argmin}} \sum_i \|\mathbf{A}_j\|_* + \frac{\mu_{i,j}}{2}\|\mathbf{X}_{i,j}^A - \mathbf{A}_j\|_F^2 + \langle \mathbf{Y}_{i,j}, \mathbf{X}_{i,j}^A - \mathbf{A}_j \rangle \tag{3.83}$$

For finding a solution, the singular value thresholding algorithm [Cai et al., 2008, Wen et al., 2011] can be used:

$$\mathbf{U}\mathbf{\Sigma}\mathbf{V}^T \quad \leftarrow \quad \frac{\sum_i \mu_{i,j}\mathbf{X}_{i,j}^A + \mathbf{Y}_{i,j}}{\sum_i \mu_{i,j}} \tag{3.84}$$

$$\mathbf{A}_j \quad = \quad \mathbf{U}S_\tau(\mathbf{\Sigma})\mathbf{V}^T \tag{3.85}$$

with $\mathbf{X}_{i,j}^A = \mathbf{X}_{i,j} - \mathbf{C}_i - \mathbf{E}_{i,j}$ and $\tau = \frac{N}{\sum_i \mu_{i,j}}$.

Sub-solution 3: The solution to the problem of finding optimal $\mathbf{C}_i$

$$\underset{\mathbf{C}_i}{\operatorname{argmin}} \frac{\mu_{i,j}}{2} \sum_j \|\mathbf{X}_{i,j}^C - \mathbf{C}_i\|_F^2 + < \mathbf{Y}_{i,j}, \mathbf{X}_{i,j}^C - \mathbf{C}_i >, \tag{3.86}$$

where $\mathbf{X}_{i,j}^C = \mathbf{X}_{i,j} - \mathbf{A}_j - \mathbf{E}_{i,j}$, can be obtained directly (by taking derivatives of the objective function and setting to zero) as

$$\mathbf{C}_i = \frac{\sum_j \mathbf{Y}_{i,j} + \mu_{i,j}\mathbf{X}_{i,j}^C}{\sum_j \mu_{i,j}}. \tag{3.87}$$

For Eq. 3.79, it was assumed that the images are properly organized so that the j_{th} instance of all classes have similar global variations, i.e., $\mathbf{A}_{i,j} = \mathbf{A}_{k,j} \ \forall \ j$ and $k \neq i$. This can be done by allowing some images to be missing for some classes. Let Ω be the set of (i, j) where $\mathbf{X}_{i,j}$ is available and $\bar{\Omega}$ is the complement of Ω. To deal with those missing entries, $\mathbf{Y}_{i,j}$, $\mu_{i,j}$ and $\mathbf{X}_{i,j}$ are set to 0 for $(i, j) \in \bar{\Omega}$ in the initialization stage. In each iteration, $\mathbf{E}_{i,j}$ for $(i, j) \in \bar{\Omega}$ is not updated. The proposed decomposition algorithm will automatically infer the missing images. For convergence, $\frac{\sum_{i,j} \|\mathbf{X}_{i,j} - \mathbf{C}_i - \mathbf{A}_j - \mathbf{E}_{i,j}\|_F^2}{\sum_{i,j}\|\mathbf{X}_{i,j}\|_F^2}$ is checked and if it is small enough (e.g., 10^{-6}), the algorithm is terminated. λ, τ and ρ are three parameters specified as input.

A similar idea was discussed in Taheri et al. [2013] to jointly perform face recognition and expression recognition, where the expression component is assumed to be superimposed onto a

Algorithm 13 The algorithm for joint sparsity model with matrix completion [Zhang and Li, 2012]

***

Input: $\mathbb{X}$, Ω, N, M, ρ, λ and τ;
Output: $\{\mathbf{C}_i\}_{i=1}^N$, $\{\mathbf{A}_j\}_{j=1}^K$ and $\{\mathbf{E}_{i,j}\}_{i,j=1}^{N,M}$;

1: $t = 0$, $\mathbf{C}_i^0 = \mathbf{A}_j^0 = \mathbf{E}_{i,j}^0 = 0$;
2: $\mathbf{Y}_{i,j}^0 = \frac{\mathbf{X}_{i,j}}{\|\mathbf{X}_{i,j}\|_F}$, $\mu_{i,j}^0 = \frac{\tau}{\|\mathbf{X}_{i,j}\|_F}$ for $(i, j) \in \Omega$;
3: $\mathbf{Y}_{i,j}^0 = 0$, $\mu_{i,j}^0 = 0$ for $(i, j) \notin \Omega$;
4: **while** not converged **do**
5:      Solve $\mathbf{E}_{i,j}$ for $(i, j) \in \Omega$ by Sub-solution 1:
6:      Solve $\mathbf{A}_j$ for $j = 1, 2, ..., M$ with Sub-solution 2;
7:      Solve $\mathbf{C}_i$ for $i = 1, 2, ..., N$ using Sub-solution 3;
8:      %Update $\mathbf{Y}_{i,j}$ and $\mu_{i,j}$ for $(i, j) \in \Omega$:
9:      $\mathbf{Y}_{i,j}^{t+1} = \mathbf{Y}_{i,j}^t + \mu_{i,j}^t(\mathbf{X}_{i,j} - \mathbf{C}_i^{t+1} - \mathbf{A}_j^{t+1} - \mathbf{E}_{i,j}^{t+1})$;
10:     $\mu_{i,j}^{t+1} = \mu_{i,j}^t \rho$;
11:     $t = t + 1$;
12: **end while**

***

expression-neural face image. In Kong and Wang [2012] a dictionary learning algorithm, DL-COPAR, was proposed for learning class-specific feature dictionaries (PARticular) and the common pattern pool (COmmon). The problem can be formulated as:

$$\{\mathbf{D}_i\}_{i=1}^{C+1}, \alpha = \min_{\{\mathbf{D}_i\}_{i=1}^{C+1}, \alpha} \sum_i^C \sum_{j \in \mathbb{I}_i} \|\mathbf{x}_j - \mathbf{D}\alpha_j\|_2^2 + \lambda\|\alpha\|_1 + \|\mathbf{x}_j - \mathbf{D}_i\alpha_j^i - \mathbf{D}_{C+1}\alpha_j^{C+1}\|_2^2,$$

(3.88)

where C is the number of categories, $\mathbb{I}_i$ is the set of data of Class i, $\mathbf{D} = [\mathbf{D}_1, \mathbf{D}_2, \cdots, \mathbf{D}_{C+1}]$ with $\mathbf{D}_1, \mathbf{D}_2, \cdots, \mathbf{D}_C$ the class specific dictionaries for class $1, \cdots, C$, respectively while $\mathbf{D}_{C+1}$ is the common dictionary and α_j^i is the subset of coefficient for sub-dictionary $\mathbf{D}_i$. By setting the selection operator $\mathbf{Q}_i$ such that $\mathbf{D}_i = \mathbf{D}\mathbf{Q}_i$ and $\alpha_j^i = \mathbf{Q}_i^T\alpha_j$, we can have:

$$\{\mathbf{D}_i\}_{i=1}^{C+1}, \alpha = \min_{\{\mathbf{D}_i\}_{i=1}^{C+1}, \alpha} \sum_i^C \sum_{j \in \mathbb{I}_i} \|\mathbf{x}_j - \mathbf{D}\alpha_j\|_2^2 + \lambda\|\alpha\|_1 + \left\|\mathbf{x}_j - \mathbf{D}\begin{bmatrix}\mathbf{Q}_i \\ \mathbf{Q}_{C+1}\end{bmatrix}\begin{bmatrix}\mathbf{Q}_i^T & \mathbf{Q}_{C+1}^T\end{bmatrix}\alpha_j\right\|_2^2.$$

(3.89)

In addition, the sparse coefficient corresponding to mis-matching particular dictionaries is required to be small, i.e., $\|(\mathbf{I} - \mathbf{Q}_i^T)\alpha_j\|_2^2$ should be small for $j \in \mathbb{I}_i$. Furthermore, the particular dictionaries were required to be incoherent, i.e., $\|\mathbf{D}_i^T \mathbf{D}_k\|_F^2$ for $i \neq k$ should be small. Thus the final formulation can be written as:

$$
\begin{aligned}
\{\mathbf{D}_i\}_{i=1}^{C+1}, \alpha \;=\; & \min_{\{\mathbf{D}_i\}_{i=1}^{C+1}, \alpha} \sum_i^C \sum_{j \in \mathbb{I}_i} \|\mathbf{x}_j - \mathbf{D}\alpha_j\|_2^2 + \left\|\mathbf{x}_j - \mathbf{D}\begin{bmatrix} \mathbf{Q}_i \\ \mathbf{Q}_{C+1} \end{bmatrix}\begin{bmatrix} \mathbf{Q}_i^T & \mathbf{Q}_{C+1}^T \end{bmatrix}\alpha_j\right\|_2^2 \\
& + \;\; \|(\mathbf{I} - \mathbf{Q}_j^T)\alpha_i\|_2^2 + \lambda\|\alpha\|_1 + \eta \sum_{k \neq i} \|\mathbf{D}_i^T \mathbf{D}_k\|_F^2 .
\end{aligned}
\tag{3.90}
$$

To solve the problem in Eq. 3.90, block coordinate descent was used, which iteratively update the sparse coefficient α, particular dictionaries $\mathbf{D}_i$ and common dictionaries $\mathbf{D}_{C+1}$. Experiments on face recognition, hand digit recognition and object recognition showed that learning shared dictionary and class-specific dictionaries enhanced the discriminative capability of the dictionaries and accordingly resulted in improved accuracy.

This idea was further explored in Zhou et al. [2012b], where a new term was introduced to enhance the discriminative capability of the dictionary. The discriminate term computes the Fisher criterion on the sparse coefficients. The method was evaluated on the ImageNet and Oxford flower datasets for image classification, with improved performance reported.

A separable dictionary learning (SeDiL) algorithm was proposed in Hawe et al. [2013]. Being separable, the dictionary $\mathbf{D}$ should be decomposed as $\mathbf{D} = \mathbf{A} \otimes \mathbf{B}$, where $\otimes$ is Kronecker product. The training data are 2-D matrices, i.e., $\mathbf{X}_i \in \mathbb{R}^{h \times w} \; \forall \; i$, which with the separable dictionary can be sparsely represented as $\mathbf{X}_i = \mathbf{A}\alpha_i \mathbf{B}^T$. The learned separable dictionary is required to be incoherent, where the incoherency $\mu(\mathbf{D})$ of a separable dictionary can be computed as

$$
\mu(\mathbf{D}) = \max\left[\mu(\mathbf{A}), \mu(\mathbf{B})\right] \leq C_1 \max\left[r(\mathbf{A}), r(\mathbf{B})\right] \leq C_2[r(\mathbf{A}) + r(\mathbf{B})]
\tag{3.91}
$$

with C_1, C_2 being some constants and $r(\cdot)$ the rank of a matrix. Formally, the problem of learning separable dictionary can be written as:

$$
\mathbf{A}, \mathbf{B}, \alpha : \min_{\mathbf{A}, \mathbf{B}, \alpha} \sum_j \frac{1}{2} \|\mathbf{A}\alpha_j \mathbf{B}^T - \mathbf{X}_j\|_F^2 + \lambda g(\alpha) + k r(\mathbf{A}) + k r(\mathbf{B}),
\tag{3.92}
$$

where $g(\alpha)$ is a sparsity measurement, and, to make the problem smooth, $g(\alpha) = \sum_i \sum_j \sum_k \log(1 + \rho\|\alpha_{ijk}\|^2)$. A geometric conjugate gradient method was used to find a solution to the above problem, which led to superlinear rate of convergence. A comparison between the dictionary learned by the K-SVD algorithm and that learned by SeDiL from 40,000 image patches is shown in Fig. 3.5(a) and (b) accordingly, with the latter showing more visible structures.

3.3.3   LEARNING DICTIONARIES FROM MULTIPLE DOMAINS

The previous two subsections discussed algorithms for learning multiple dictionaries from data of one domain (e.g., one source). There are applications where data of different domains may need to be processed at the same time. One example is data of multiple modalities. This subsection

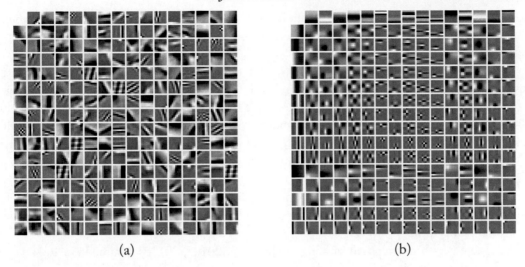

(a)   (b)

Figure 3.5: A comparison between the dictionary learned by the K-SVD algorithm (a) and dictionary learned by SeDiL (b). The dictionary is learned from 40,000 8×8 patches. For SeDiL, the dictionary $\mathbf{A} \otimes \mathbf{B}$ is shown. The figure is from Fig. 1 of Hawe et al. [2013].

introduces algorithms for learning multiple dictionaries in such applications. One unique aspect of this problem is that there may be additional cross-domain information that can be or should be captured by the learning process.

In Yang et al. [2008, 2010a], two dictionaries are learned jointly, one based on patches from high-resolution images (referred to as the high-resolution dictionary $\mathbf{D}_H$) and the other based on patches of low-resolution images (referred to as low-resolution dictionary $\mathbf{D}_L$). The one-to-one correspondence between the patches of high-resolution images and patches of low-resolution images are assumed known. A patch of a high-resolution image $\mathbf{x}^H$ and its corresponding patch of a low-resolution image $\mathbf{x}^L$ (referred to as patch pair $(\mathbf{x}^H, \mathbf{x}^L)$) are required to share the same sparse coefficient α under the high-resolution dictionary $\mathbf{D}^H$ and the low-resolution dictionary $\mathbf{D}^L$ accordingly. Formally, the problem can be written as:

$$\mathbf{D}^H, \mathbf{D}^L, \alpha \quad : \quad \min_{\mathbf{D}^H, \mathbf{D}^L, \alpha} \sum_i \frac{1}{2}\|\mathbf{x}_i^H - \mathbf{D}^H \alpha_i\|_2^2 + \frac{1}{2}\|\mathbf{x}_i^L - \mathbf{D}^L \alpha_i\|_2^2 + \lambda\|\alpha_i\|_1 \quad (3.93)$$

$$\min_{\mathbf{D}^H, \mathbf{D}^L, \alpha} \sum_i \frac{1}{2}\left\| \begin{bmatrix} \mathbf{x}_i^H \\ \mathbf{x}_i^L \end{bmatrix} - \begin{bmatrix} \mathbf{D}^H \\ \mathbf{D}^L \end{bmatrix} \alpha_i \right\|_2^2 + \lambda\|\alpha_i\|_1. \quad (3.94)$$

The above problem can be solved via a modified K-SVD algorithm. From Eq. 3.94, we can find that, there is a one-to-one correspondence from the atoms of the high-resolution dictionary $\mathbf{D}^H$ and those of the low-resolution dictionary $\mathbf{D}^L$. Then given a new low-resolution patch $\mathbf{x}^L$, we can compute its corresponding high-resolution patch as: $\mathbf{x}^H = \mathbf{D}^H \alpha$, where α : $\min_\alpha \frac{1}{2} \|\mathbf{x}^L - \mathbf{D}^L \alpha\|_2^2 + \lambda \|\alpha\|_1$, which was used in Yang et al. [2008] for single-image super-resolution.

This idea was further explored in Yang et al. [2011b], where the low-resolution patches are grouped in several groups via K-means and then dictionary learning is applied to patches of each group independently to generate multiple high-resolution dictionary and low-resolution dictionary pairs. A multi-task learning framework was used for learning multiple high-resolution and low-resolution dictionary pairs. This resulted in a more efficient super resolution algorithm, since sparse coding can be done much more efficiently by only considering the atoms from the dictionary of one low-resolution path group. In Liu et al. [2013c], this idea was extended for person re-identification across views. Local coordinate code (LCC) was introduced to enhance the discriminative capability of the dictionary and formulated under a semi-supervised dictionary learning framework.

The algorithms proposed in Yang et al. [2008, 2010a, 2011b] assumed one-to-one correspondence between the atoms of the high-resolution dictionary and those of the low-resolution dictionary, which could be too restrictive in some scenarios. An improved method was proposed in Wang et al. [2012], where the atoms of the two dictionaries are semi-coupled via a linear projection matrix, which provides more flexibility. The algorithm, named semi-coupled dictionary learning, can be formulated as:

$$\mathbf{D}^x, \mathbf{D}^y, \mathbf{W}, \alpha^x, \alpha^y \quad : \quad \min_{\mathbf{D}^x, \mathbf{D}^y, \mathbf{W}, \alpha^x, \alpha^y} \|\mathbf{X} - \mathbf{D}^x \alpha^x\|_F^2 + \|\mathbf{Y} - \mathbf{D}^y \alpha^y\|_F^2 \tag{3.95}$$
$$+ \gamma \|\alpha^y - \mathbf{W}\alpha^x\|_F^2 + \lambda^x \|\alpha^x\|_1 + \lambda^y \|\alpha^y\|_1 + \lambda^w \|\mathbf{W}\|_F^2,$$
$$\text{s.t.} \quad \|\mathbf{d}_i^x\|_2 \leq 1, \|\mathbf{d}_i^y\|_2 \leq 1 \; \forall i$$

where $\mathbf{W}$ is the linear projection matrix which semi couples the atoms of the two dictionaries $\mathbf{D}^x$ and $\mathbf{D}^y$ (and thus sparse coefficients α^x and α^y).

The problem in Eq. 3.95 can be solved by the block coordinate descent algorithm. When solving α^x and $\mathbf{D}^x$ with other variables fixed, we have the following problems:

$$\mathbf{D}^x, \alpha^x \quad : \quad \min_{\mathbf{D}^x, \alpha^x} \|\mathbf{X} - \mathbf{D}^x \alpha^x\|_F^2 + \|\alpha^y - \mathbf{W}\alpha^x\|_F^2 + \lambda^x \|\alpha^x\|_1 \tag{3.96}$$

$$\min_{\mathbf{D}^y, \alpha^x} \| \begin{bmatrix} \mathbf{X} \\ \alpha^x \end{bmatrix} - \begin{bmatrix} \mathbf{D}^x \\ \mathbf{W} \end{bmatrix} \alpha^x \|_F^2 + \lambda^x \|\alpha^x\|_1 \tag{3.97}$$

which can be solved via a K-SVD-like algorithm. α^y and $\mathbf{D}^y$ can be solved similarly. Considering $\mathbf{W}$, we have $\mathbf{W} : \min_{\mathbf{W}} \gamma \|\alpha^y - \mathbf{W}\alpha^x\|_F^2 + \lambda^y \|\mathbf{W}\|_F^2$, which indicates a closed form solution $\mathbf{W} = \alpha^y (\alpha^x)^T [\alpha^x (\alpha^x)^T + \frac{\lambda^w}{\gamma}]^{-1}$. Due to the flexibility introduced by the linear projection matrix $\mathbf{W}$, the method proposed in Wang et al. [2012] can be used not only for image super-resolution as Yang et al. [2008, 2010a, 2011b], but also for face image synthesis between sketch and photo (e.g., see Fig. 5 of Wang et al. [2012] for an example).

In Huang and Wang [2013], instead of requiring the sparse coefficients in different domains to be identical (e.g., $\alpha_i^x = \alpha_i^y$) or coupled via a linear projection matrix (i.e., $\alpha_i^y = \mathbf{W}\alpha_i^x$), it is assumed that the sparse coefficients of different domains can be projected onto an identical vector, i.e., $\mathbf{W}^x \alpha_i^x = \mathbf{W}^y \alpha_i^y$, hence leading to the following problem:

$$
\mathbf{D}^x, \mathbf{D}^y, \mathbf{W}^x, \mathbf{W}^y, \alpha^x, \alpha^y \quad : \quad \min_{\mathbf{D}^x, \mathbf{D}^y, \mathbf{W}^x, \mathbf{W}^y, \alpha^x, \alpha^y} \|\mathbf{X} - \mathbf{D}^x \alpha^x\|_F^2 + \|\mathbf{Y} - \mathbf{D}^y \alpha^y\|_F^2 \tag{3.98}
$$
$$
+ \gamma \|\mathbf{W}^y \alpha^y - \mathbf{W}^x \alpha^x\|_F^2 + \lambda^x \|\alpha^x\|_1 + \lambda^y \|\alpha^y\|_1 + \lambda^w \|\mathbf{W}\|_F^2
$$
$$
+ \lambda^r \|(\mathbf{W}^x)^{-1}\|_1 + \lambda^r \|(\mathbf{W}^y)^{-1}\|_1
$$
$$
\text{s.t.} \quad \|\mathbf{d}_i^x\|_2 \le 1, \|\mathbf{d}_i^y\|_2 \le 1 \; \forall i,
$$

where the constraint $\lambda^r \|(\mathbf{W}^x)^{-1}\|_1 + \lambda^r \|(\mathbf{W}^y)^{-1}\|_1$ is introduced to avoid the trivial solution like $\mathbf{W}^x = \mathbf{W}^y = 0$. Block coordinate descent can again be used to solve the problem.

In Zheng et al. [2012], the idea proposed in Yang et al. [2008, 2010a, 2011b] was studied under a supervised learning framework, where discriminative learning was introduced to Eq. 3.94. The learned dictionary was applied in cross-view action recognition, where two dictionaries are learned from two views. A more general problem was considered in Qiu et al. [2012], where more than two views (or domains) were allowed. In particular, the dictionary $\mathbf{D}$ is represented as a parametric function $\mathbf{D} = F(\theta, \mathbf{W})$, where θ is the domain parameter and $\mathbf{W}$ is the dictionary function parameter. Accordingly, dictionary learning can be formulated as:

$$
\theta, \mathbf{W}, \alpha \quad : \quad \min_{\theta, \mathbf{W}, \alpha} \sum_i \|\mathbf{X}_i - F(\theta_i, \mathbf{W})\alpha\|_F^2 \tag{3.99}
$$

$$
\min_{\theta, \mathbf{W}, \alpha} \left\| \begin{bmatrix} \mathbf{X}_1 \\ \mathbf{X}_2 \\ \cdots \\ \mathbf{X}_N \end{bmatrix} - \begin{bmatrix} F(\theta_1, \mathbf{W}) \\ F(\theta_2, \mathbf{W}) \\ \cdots \\ F(\theta_N, \mathbf{W}) \end{bmatrix} \alpha \right\|_F^2 \text{ s.t. } \|\alpha_i\|_0 \le T.
$$

Equation 3.99 can be solved via block coordinate descent. First, the dictionaries $\mathbf{D}_i = F(\theta_i, \mathbf{W})$ are computed with the K-SVD algorithm; then the domain parameter θ_i and dictionary function parameter $\mathbf{W}$ are updated via gradient descent. The learned parameterized dictionaries can be applied in classification cross domains. For a test input, the domain parameter θ is first computed via binary search; then the domain specific dictionary is generated by using the computed domain parameter; and finally classification is applied on the generated dictionary. The experiment on the CMU-PIE dataset showed that the method was able to achieve good results in the presence of illumination and pose variations. An example of transferring face image cross different poses (pose synthesis) on CMU-PIE dataset can be found in Fig. 6 of Qiu et al. [2012]. Joint dictionary learning was also proposed for biometric recognition in Shekhar et al. [2014], where multiple modalities of biometric feature were combined and yielded better performances than using single modality or existing fusion-based approaches.

The performance of the parameterized dictionary proposed in Zheng et al. [2012] depends on selection of the dictionary parameter function. To achieve better performance, more complex parameter functions (e.g., higher-order polynomials) should be used, which would, however, result in larger computational cost. In Ni et al. [2013], this problem was tackled via learning a set of dictionaries for some intermediate domain (referred to as intermediate dictionaries), which is termed subspace interpolation. Starting from the dictionary $\mathbf{D}^0$ of the source domain, a set of intermediate dictionaries $\{\mathbf{D}^k\}_{k=1}^K$ are learned to approach the target domain $\mathbf{X}^t$. Specifically, given the dictionary $\mathbf{D}^k$ of the k_{th} domain, the next intermediate dictionary $\mathbf{D}^{k+1}$ can be learned as: $\mathbf{D}^{k+1} = \mathbf{D}^k + \Delta\mathbf{D}^k$, with $\Delta\mathbf{D}^k$ computed as:

$$\Delta\mathbf{D}^k : \min_{\Delta\mathbf{D}^k} \|\mathbf{J}^k - \Delta\mathbf{D}^k \alpha^k\|_F^2, \tag{3.100}$$

where $\mathbf{J}^k = \mathbf{X}^t - \mathbf{D}^k \alpha^k$ is the reconstruction residual from $\mathbf{D}^k$ and α^k is the sparse coefficient computed from $\alpha^k : \min_{\alpha^k} \|\mathbf{Y}^t - \mathbf{D}^k \alpha^k\|_F^2$ s.t. $\forall i \|\alpha_i^k\|_0 \le T$. The above problem has a closed form solution $\Delta\mathbf{D}^k = \mathbf{J}^k (\alpha^K)^T (\lambda + \alpha^k (\alpha^k)^T)^{-1}$. The intermediate dictionaries are generated sequentially, until $\|\Delta\mathbf{D}^k\|_F \le \delta$. The learned set of intermediate dictionaries together with the source domain dictionary $\mathbf{D}^0$ and the target domain dictionary $\mathbf{D}^K$ provide a way of extracting features that are more robust cross domains:

$$[(\mathbf{D}^0\alpha)^T, (\mathbf{D}^1\alpha)^T, \cdots, (\mathbf{D}^K\alpha)^T]^T, \tag{3.101}$$

where α is the sparse coefficient computed from either the source domain or the target domain.

3.3.4   LEARNING DICTIONARIES WITH A HIERARCHY

In many real-world problems, data may often be naturally organized in hierarchies. For example, the object categories in a large-scale image dataset may form a tag taxonomy. Hierarchical dictionaries, especially tree-structured dictionaries, have been widely used in image-related tasks, such as compression with wavelet dictionaries. This subsection presents several dictionary learning algorithms that take into consideration the potential hierarchical structure in the data.

In Jenatton et al. [2010, 2011], a hierarchical sparsity-inducing norm was used to enforce the hierarchical structure in the dictionary. Formally the problem is defined as:

$$\mathbf{D}, \alpha : \min_{\mathbf{D},\alpha} \sum_i \frac{1}{2}\|\mathbf{x}_i - \mathbf{D}\alpha_i\|_2^2 + \lambda\Omega(\alpha_i) \text{ s.t. } \|\mathbf{d}_i\|_2^2 = 1, \|\alpha_i\|_0 \le t, \tag{3.102}$$

where $\mathbf{D}$ is the hierarchical dictionary and $\Omega(\cdot)$ is the hierarchical sparsity-inducing norm. The dictionary elements are organized in a rooted tree $\mathcal{T}$, where each node p corresponds to a dictionary atom and nodes sharing the same parent node form a group of dictionary atoms. An example is shown in Fig. 3.6. To enforce such a structure, we can define a group of dictionary atoms which contains an atom and all of its descendent atoms and then penalize the number of such groups used for representation, which is induced by $\Omega(\cdot)$. A convex measurement was proposed for $\Omega(\cdot)$ [Bach, 2009, Kim and Xing, 2010, Zhao et al., 2009]:

$$\Omega\alpha = \sum_{g\in\mathcal{G}} w_g \|\alpha_g\|_p, \tag{3.103}$$

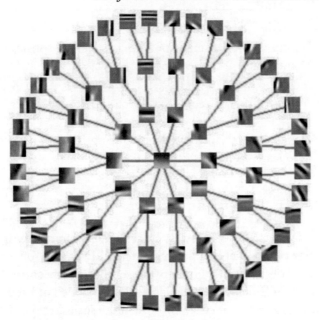

Figure 3.6: The examples of learned tree-structured dictionary. The figure is from Fig. 6 of Jenatton et al. [2010].

where $\|\cdot\|_p$ could be either ℓ_∞ or ℓ_2. A block coordinate descent method can be used to solve the problem in Eq. 3.102 in its dual form.

A two-layer dictionary learning algorithm was proposed in Yu et al. [2011], where the bottom layer encodes the local patches of an image and the top layer encodes regions or sets of local patches via pooling. The problem is formulated as:

$$\mathbf{D}, \mathbf{B}, \alpha, \beta \; : \quad \min_{\mathbf{D}, \mathbf{B}, \alpha, \beta} \sum_i \frac{1}{2} \|\mathbf{x}_i - \mathbf{B}\beta_i\|_2^2 + \lambda_2 \beta_i^T \Omega(\alpha)\beta_i + \lambda_1 \|\beta_i\|_1 + \gamma \|\alpha\|_1 \quad (3.104)$$

$$\min_{\mathbf{D}, \mathbf{B}, \alpha, \beta} \sum_i \frac{1}{2} \|\mathbf{x}_i - \mathbf{B}\beta_i\|_2^2 + \lambda_1 \|\beta_i\|_1 + \text{tr}(\beta^T \Omega(\alpha)\beta) + \gamma \|\alpha\|_1, \quad (3.105)$$

where $\mathbf{B}$ and β are, respectively, the dictionary and sparse code for the bottom layer, $\mathbf{D}$ and α are, respectively, the dictionary and sparse coefficient for the top layer, and $\Omega(\alpha) = (\sum_k \alpha_k \text{diag}(\mathbf{d}_k))^{-1}$. The above formulation encourages sparsity for both patch level and patch-set level. In $\text{tr}(\beta^T \Omega(\alpha)\beta) = \text{tr}((\beta\beta^T)\Omega(\alpha))$, $(\beta\beta^T)$ is the co-variance of the sparse coefficients of the patches. Accordingly, to minimize $\text{tr}(\beta^T \Omega(\alpha)\beta)$, we have $\Omega(\alpha) = (\beta\beta^T)^{-1}$, i.e., the set-level sparse code can be viewed as modeling the co-variance structure of the patch-level representation. A block coordinate descent method was used to solve Eq. 3.104, where the problems of updating

---

Algorithm 14 The algorithm proposed in Ophir et al. [2011]

Input : input image $\mathbf{X}$, number of decomposition level s, sparsity constraint T
Output : dictionary $\mathbf{D}$ and sparse coefficient α

1: Apply 2D wavelet transform to decompose all training image into $3s + 1$ bands
2: **for** each band b **do**
3:     Extract the patches with maximal overlapping from Band b of all training images
4:     Apply K-SVD to learn dictionary of Band b
5: **end for**

---

the sparse coefficient for patch-level β and patch-set-level α can be rewritten as a LASSO problem. The dictionary for patch-level $\mathbf{B}$ and patch-set-level $\mathbf{D}$ can be solved via gradient method.

Learning dictionaries with more than two levels has also been studied, where the earliest work can be traced back to Etemad and Chellappa [1998], which utilized wavelet packet to learn a discriminative dictionary. In Ophir et al. [2011], a multi-scale dictionary learning algorithm was proposed, which combines the power of wavelet analysis and the K-SVD algorithm. The learned dictionary is able to deliver multi-scale information of the signal. By using the wavelet synthesis operator (or inverse wavelet transform), the problem can be written as:

$$\mathbf{D}, \alpha : \min_{\mathbf{D},\alpha} \|\mathbf{X} - \mathbf{W}_s\mathbf{D}\alpha\|_F^2 \text{ s.t. } \|\alpha_i\|_0 \leq T \ \forall i. \tag{3.106}$$

Assuming that the wavelet synthesis operator $\mathbf{W}_s$ is square and unitary (i.e., orthogonal wavelet with periodic extension), we can utilize the wavelet analysis operator $\mathbf{W}_a$, where $\mathbf{W}_a\mathbf{W}_s = \mathbf{I}$, to rewrite it as:

$$\mathbf{D}, \alpha : \min_{\mathbf{D},\alpha} \|\mathbf{W}_a\mathbf{X} - \mathbf{D}\alpha\|_F^2 \text{ s.t. } \|\alpha_i\|_0 \leq T \ \forall i. \tag{3.107}$$

Accordingly, a dictionary is learned from the wavelet-decomposed image $\mathbf{W}_a\mathbf{X}$. An example can be found in Fig. 4 of Ophir et al. [2011]. With wavelet transform, the training data is decomposed into a collection of bands at different scales and orientation. Thus, it makes senses to learn a separate dictionary for each band. Assuming that $(\mathbf{W}_a\mathbf{X})_b$ is the b_{th} band, the dictionary of the b_{th} band $\mathbf{D}_b$ can be learned as:

$$\mathbf{D}_b, \alpha_b : \min_{\mathbf{D}_b,\alpha_b} \|(\mathbf{W}_a\mathbf{X})_b - \mathbf{D}_b\alpha_b\|_F^2 \text{ s.t. } \|\alpha_{i,b}\|_0 \leq T_b \ \forall i. \tag{3.108}$$

The algorithm is summarized in Algorithm 14.

The method proposed in Ophir et al. [2011] was generalized in Xiang et al. [2011], where the wavelet decomposition operator $\mathbf{W}_a$ is replaced by random projections $\mathbf{T}_k$. Each $\mathbf{T}_k$ has orthonormal rows and the rows for different $\mathbf{T}_k$ are orthogonal. It was proven in Xiang et al.

[2011] that (Theorem 2), under mild conditions, let α_1 and α_2 be the sparse representations of the random projected data $\mathbf{Tx}_1$ and $\mathbf{Tx}_2$ on the dictionary $\mathbf{D}$, with probability $1 - \rho$:

$$\frac{p}{d}(1 - \epsilon_1)(1 - \epsilon_2)\|\alpha_1 - \alpha_2\|_2^2 \leq \|\mathbf{x}_1 - \mathbf{x}_2\|_2^2 \leq \frac{p}{d}(1 + \epsilon_1)(1 + \epsilon_2)(\|\alpha_1 - \alpha_2\|_2^2 + 2\mu_1^2 + 2\mu_2^2),$$

(3.109)

where $\mu_i = \|\mathbf{Tx}_i - \mathbf{D}\alpha_i\|_2$, ϵ_1, ϵ_2 are two small constants, and p and d are constants. This theorem shows that, even with some random projection $\mathbf{T}$, the distance of a pair of data points will be preserved when represented under some dictionary $\mathbf{D}$. Thus we can replace the wavelet analysis operator $\mathbf{W}_a$ in Ophir et al. [2011] by some random projection matrix, which has orthonormal rows and the rows for different $\mathbf{T}_k$ are orthogonal, to learn a multi-scale dictionary.

Tree K-SVD, an extension of K-SVD for learning tree-structured dictionary, was proposed in Mazaheri et al. [2013]. A L-level tree K-SVD contains K^l dictionaries of K atoms at Level l, where $l = 0$ is the root level, and each dictionary except the one at the root level is linked to a dictionary atom of its parent node. The process of sparse coding a signal with a dictionary learned by tree K-SVD can be modeled as finding a path from the root node to a leaf node (node at Level $L - 1$). Each node is a dictionary at certain level and only one dictionary atom can be selected for each of the dictionaries (i.e., sparsity constraint for each dictionary is 1). One advantage of such a sparse coding process is that, the sparisity of the coefficient can be easily controlled by the levels of dictionary traversed and more important atoms are found earlier from dictionaries closer to the root level. This process can be done iteratively as illustrated in Algorithm 15. The Tree K-SVD dictionary can be learned in a similar way, which is summarized in Algorithm 16.

However, requiring only one atom to be used for each dictionary can be too restrictive. For example, the final number of atoms used for sparse representation is limited by the number of levels traversed, while going too deep will make the size of the tree K-SVD dictionary explode. In addition there will be much less data available for training dictionary at deeper levels. To this end, adaptive sparse coding was also proposed in Mazaheri et al. [2013], where selecting more than 1 dictionary atom from each dictionary is allowed. More specifically, whether or not to go to the next level will be decided based on the reconstruction error. As a result, the sparsity of the level and number of levels can be adapted to minimize the distortion.

In Shen et al. [2013], multi-level dictionary learning is combined with discriminative dictionary learning for learning a multi-scale discriminative dictionary. To adapt to the tag taxonomy of the images, the multi-level dictionary is also organized into a tree, where each node, which is attached to a tag, is associated with a dictionary and a classifier. Accordingly, the classification process is equivalent to finding the optimal path from the root node to a leaf node. Each internal node corresponds to a multi-class classification problem, and assigns the input to one of its child nodes. The dictionary of each node, except for the root node, consists of two parts: the dictionary inherited from its parent node and the dictionary unique to itself. The multi-level dictionary can be learned layer-by-layer, starting from the root level, as in Mazaheri et al. [2013]. A similar idea has also been studied in Zheng and Jiang [2013], where a level-specific dictionary was introduced, which is the concatenation of dictionaries of the nodes in the same level, and

---

Algorithm 15 The sparse coding algorithm for Tree K-SVD

Input : input signal, dictionary learned by tree K-SVD, maximal level for sparse coding
Output : sparse coefficient

1: Starting at root node
2: **while** maximal level of sparse coding not reached **do**
3:    Apply OMP with sparsity constraint equal to 1 to find the sparse coefficient with current dictionary
4:    Compute the signal residual with current sparse coefficient
5:    Go to the node which is linked to the only dictionary atom used for sparse representation
6: **end while**

Algorithm 16 The dictionary learning algorithm for Tree K-SVD

Input : input signal, maximal level for sparse coding
Output : tree K-SVD dictionary

1: Starting at root node
2: Learn the dictionary and sparse coefficient via K-SVD with sparsity constraint set to 1
3: Compute the signal residuals with current sparse coefficient
4: Cluster the signal residuals into K groups according to their usage of dictionary atoms
5: **if** maximal level of sparse coding reached **then**
6:    Terminate
7: **end if**
8: Learn a dictionary for each group of sparse residuals by repeating the above steps and set its parent to the corresponding dictionary atom

---

a level-specific classifier is also learned for each level-specific dictionary. In Yang et al. [2013b], a multi-scale dictionary is learned from the decomposed multi-scale ridgelet support vectors of natural images for image denoising, where the learned dictionary is able to preserve the edges, contour, textures in the images and avoid the ringing effects.

3.4   ONLINE DICTIONARY LEARNING

In the previous sections, we described various dictionary learning algorithms that typically assume the availability of a set of training data prior to learning. Once the dictionary is learned, it is fixed for later use. This is essentially a batch-processing method, which may not be appropriate for some applications where the training data may either be too big to fit into the memory in a single batch, or come into the system as live streams and we need to have the dictionary incrementally updated based on newly available data items. In this section, we describe dictionary learning algorithms that are capable of tackling incremental or on-line learning. We will refer to these algorithms as online dictionary learning algorithms.

One of the earliest online dictionary learning algorithms was proposed in Mairal et al. [2009a], which formulates the dictionary learning problem under an online learning framework via stochastic approximation. Recall that the dictionary learning problem can be written as:

$$\mathbf{D}, \alpha : \min_{\mathbf{D}, \alpha} \frac{1}{2}\|\mathbf{X} - \mathbf{D}\alpha\|_F^2 + \|\alpha\|_1 \text{ s.t. } \|\mathbf{d}_i\|_2 = 1 \forall i. \tag{3.110}$$

The problem related to updating $\mathbf{D}$ given data $\mathbf{x}_1, \mathbf{x}_2, \cdots, \mathbf{x}_t$ can be written as

$$\mathbf{D} : \min_{\mathbf{D}^t} \operatorname{tr}(\mathbf{D}^T \mathbf{A}_t) - \operatorname{tr}(\mathbf{D}^T \mathbf{B}_t), \tag{3.111}$$

where $\mathbf{A}_t = \frac{1}{2}\sum_i^t \alpha_i \alpha_i^T$ and $\mathbf{B}_t = \frac{1}{2}\sum_i^t \mathbf{x}_i \alpha_i^T$. By some manipulations of the terms, we can obtain the solution for $\mathbf{D}$ as

$$\mathbf{u}_j = \frac{\mathbf{b}_j - \mathbf{D}\mathbf{a}_j}{\mathbf{A}_j} + \mathbf{d}_j \tag{3.112}$$
$$\mathbf{d}_j = \frac{\mathbf{u}_j}{\max\left[\|\mathbf{u}_j\|_2, 1\right]}$$

with $\mathbf{d}_j, \mathbf{b}_j, \mathbf{a}_j$ are the j_{th} column of the dictionary $\mathbf{D}, \mathbf{A},$ and $\mathbf{B}$, respectively. To develop an on-line version of the above procedure, we only need to develop an incremental updating scheme for $\mathbf{A}_t$ and $\mathbf{B}_t$, which is given below:

$$\mathbf{A}_t = \beta \mathbf{A}_{t-1} + \frac{1}{2}\sum_i^\eta \alpha_{t,i} \alpha_{t,i}^T \tag{3.113}$$
$$\mathbf{B}_t = \beta \mathbf{B}_{t-1} + \frac{1}{2}\sum_i^\eta \mathbf{x}_{t,i} \alpha_{t,i}^T,$$

where η is the size of the batch (which can be as small as 1), $\beta = \frac{\theta+1-\eta}{\theta+1}$, and $\mathbf{x}_{t,i}$ is the i_{th} data sampled from the t_{th} epoch, with $\theta = t\eta$, if $t < \eta$, or $\theta = \eta^2 + t - \eta$, if $t \geq \eta$. The full algorithm is summarized in Algorithm 17. To obtain the initial dictionary $\mathbf{D}_0$, one can simply use the data from the first several batches. The proof of convergence of the algorithm was also presented in Mairal et al. [2009a].

A similar idea was studied in Skretting and Engan [2010, 2011], where recursive least square was utilized in updating the dictionary. By using least square, the dictionary is updated as $\mathbf{D}_t = \mathbf{B}_t \mathbf{A}_t^{-1}$. However, $\mathbf{A}_t$ is updated differently, by writing $\mathbf{C}_t = \mathbf{A}_t^{-1}$, $\mathbf{C}_{t+1} =$

Algorithm 17 The online dictionary learning algorithm proposed in Mairal et al. [2009a]

Input : the stream of input data $p(x)$, initial dictionary $\mathbf{D}_0$
Output : dictionary $\mathbf{D}$

1: **for** $t = 1, \cdots, T$ **do**
2:     Draw a mini-batch of data $\mathbf{x}_{t,1}, \mathbf{x}_{t,2}, \cdots, \mathbf{x}_{t,\eta}$ from stream
3:     Apply sparse coding on the new data to have the coefficient $\alpha_{t,1}, \alpha_{t,2}, \cdots, \alpha_{t,\eta}$
4:     Update the statistics $\mathbf{A}_t$ and $\mathbf{B}_t$ with Eq. 3.113
5:     Update the dictionary $\mathbf{D}_t$ atom by atom with Eq. 3.112
6: **end for**

$\mathbf{C}_t - \frac{\mathbf{C}_t \alpha_t \alpha_t^T \mathbf{C}_t}{\alpha_t^T \mathbf{C}_t \alpha_t + 1}$. Accordingly, we have:

$$
\begin{aligned}
\mathbf{D}_t &= \mathbf{B}_t \mathbf{A}_t^{-1} & (3.114) \\
&= \mathbf{B}_t \mathbf{C}_t \\
&= (\mathbf{B}_{t-1} + \mathbf{x}_t \alpha_t^T)(\mathbf{C}_{t-1} - \frac{\mathbf{C}_{t-1} \alpha_t \alpha_t^T \mathbf{C}_{t-1}}{\alpha_t^T \mathbf{C}_{t-1} \alpha_t}) \\
&= \mathbf{B}_{t-1} \mathbf{C}_{t-1} - \mathbf{B}_{t-1} \frac{\mathbf{C}_{t-1} \alpha_t \alpha_t^T \mathbf{C}_{t-1}}{\alpha_t^T \mathbf{C}_{t-1} \alpha_t} + \mathbf{x}_t \alpha_t^T \mathbf{C}_{t-1} - \mathbf{x}_t \alpha_t^T \frac{\mathbf{C}_{t-1} \alpha_t \alpha_t^T \mathbf{C}_{t-1}}{\alpha_t^T \mathbf{C}_{t-1} \alpha_t + 1} \\
&= \mathbf{D}_{t-1} + \frac{\mathbf{x}_t - \mathbf{D}_{t-1} \alpha_t}{1 + \alpha_t^T \mathbf{C}_{t-1} \alpha_t} \alpha_t^T \mathbf{C}_{t-1}.
\end{aligned}
$$

To improve the convergence behavior and reduce the effect of the initialization point, a forgetting parameter $0 \leq \lambda \leq 1$ was introduced, which casts smaller weight on earlier data and larger weight on newer data. A small value is initially used for λ, which gradually increases to 1. Therefore the initially random (and presumably not very good) dictionary will be quickly forgotten, and as λ increases, the dictionary is remembered. Several choices for λ, e.g., linear, quadratic, cubic, hyperbola and exponential, were described in Skretting and Engan [2010]. The problem of online dictionary learning with forgetting parameter can be written as:

$$
\mathbf{D}, \alpha : \min_{\mathbf{D}, \alpha} \frac{1}{2} \sum_{i}^{t} \lambda^{t-i} \|\mathbf{x}_i - \mathbf{D}\alpha_i\|_2^2 + \|\alpha_i\|_1 \text{ s.t. } \|\mathbf{d}_i\|_2 = 1 \forall i. \tag{3.115}
$$

Accordingly, the algorithm can be revised as:

$$\mathbf{A}_t = \lambda\mathbf{A}_{t-1} + \frac{1}{2}\sum_i^{\eta} \alpha_{t,i}\alpha_{t,i}^T \qquad (3.116)$$

$$\mathbf{B}_t = \lambda\mathbf{B}_{t-1} + \frac{1}{2}\sum_i^{\eta} \mathbf{x}_{t,i}\alpha_{t,i}^T$$

$$\mathbf{D}_t = \mathbf{D}_{t-1} + \frac{1}{\lambda}\frac{\mathbf{x}_t - \mathbf{D}_{t-1}\alpha_t}{1 + \alpha_t^T\mathbf{C}_{t-1}\alpha_t}\alpha_t^T\mathbf{C}_{t-1}.$$

The idea was further studied in Lu et al. [2013], where the ℓ_1 norm instead of the ℓ_2 norm was used for the reconstruction error. As a result, the method proposed in Lu et al. [2013] is more robust to outliers than those proposed in Mairal et al. [2009a] and Skretting and Engan [2010, 2011]. The objective function is written as:

$$\mathbf{D}, \alpha : \min_{\mathbf{D},\alpha} \|\mathbf{X} - \mathbf{D}\alpha\|_1 + \|\alpha\|_1 \text{ s.t. } \|\mathbf{d}_i\|_2 = 1 \forall i. \qquad (3.117)$$

The dictionary is updated dimension-by-dimension, instead of atom-by-atom:

$$\mathbf{d}^j : \min_{\mathbf{d}^j} \sum_{i=1}^{t} \|x_i^j - \mathbf{d}^j\alpha_i\|_1 \qquad (3.118)$$

which can be solved by iteratively solving the following problem until convergence:

$$\mathbf{d}^j : \min_{d_j} \sum_{i=1}^{t} w_i^j (x_i^j - \mathbf{d}^j\alpha_i)^2 \rightarrow \mathbf{d}^j = \sum_{i=1}^{t} w_i^j x_i^j \alpha_i (\sum_{i=1}^{t} w_i^j \alpha_i\alpha_i^T)^{-1}, \qquad (3.119)$$

where the weight w_i^j is computed as $w_i^j = \frac{1}{\sqrt{(x_i^j - \mathbf{d}^j\alpha_i)^2 + \delta}}$ with δ a small constant. Accordingly the two statistics $\mathbf{A}$ and $\mathbf{B}$ are computed as:

$$\mathbf{A}_t^j = = \sum_{i=1}^{t} w_t^j \alpha_t\alpha_t^T = \mathbf{A}_{t-1} + w_t^j \alpha_t\alpha_t^T \qquad (3.120)$$

$$\mathbf{B}_t^j = = \sum_{i=1}^{t} w_t^j x_t^j \alpha_t^T = \mathbf{B}_{t-1} + w_t^j x_t^j \alpha_t^T,$$

where $\mathbf{A}_t^j$ is the j_{th} row of $\mathbf{A}_t$. The weight w_i^j measures the quality of sparse representation for the j_{th} dimension of the training data. Comparing the method proposed in Lu et al. [2013] with those reported in Mairal et al. [2009a] and Skretting and Engan [2010, 2011], we can view that the former utilizes data-adaptive weight w_i^j to adjust the importance of data to different dimensions of the dictionary, which leads to better robustness in face of outliers.

In addition to the afore-mentioned, online dictionary learning has seen quite some applications. For example, in Xie et al. [2010], online dictionary learning was used for local coordinate coding in large-scale datasets. Online learning was also used for learning dictionaries with overlapping group structures in Szabó et al. [2011]. In Zhang et al. [2013a], an online, semi-supervised method was proposed for discriminative dictionary learning. In the following, we elaborate an example of the application of online dictionary learning on a visual computing task, visual tracking.

By combing dictionary learning with online learning, the power of sparse coding can be employed in visual tracking to achieve robustness to outliers and partial occlusion [Bai and Li, 2012]. The assumption is that, the target being tracked can be sparsely represented by its examples in the previous frames. Accordingly, a dictionary is learned from the examples in the previous frames and updated given a new observed frame, via online dictionary learning:

$$\mathbf{D}, \alpha : \min_{\mathbf{D}, \alpha} \frac{1}{2}\|\mathbf{X} - \mathbf{D}\alpha\|_F^2 + \|\alpha\|_1 \text{ s.t. } \|\mathbf{d}_i\|_2 + \lambda\|\mathbf{d}_i\|_1 \leq 1 \forall i, \qquad (3.121)$$

where $\|\mathbf{d}_i\|_2 + \lambda\|\mathbf{d}_i\|_1 \leq 1$ is called elastic-net constraint, which is able to preserve the highly correlated entries in the dictionary that can better capture local appearances of the target.

Given a new frame, some windows are sampled according to some distribution. Sparse coding is then applied for those windows, with those with high reconstruction error or violating the sparsity constraint being rejected as outliers. Estimation of the target in the current frame is done according to the sparse representation of the sampled windows. The dictionary is then updated with the new estimation of the target. And the process continues with the next frame. Such a process is summarized in Algorithm 18. The initial dictionary can be built manually or learned with windows sampled around the target in the first frames.

On the task of visual tracking, similar ideas have also been studied in Zhang et al. [2012c], which introduces the $\ell_{p,q}$ mixture norm and multi-task learning to online dictionary learning, Wang et al. [2013a], which enforces the sparse coefficient and the dictionary to be nonnegative to gain robustness over occlusions and other variations, and Xing et al. [2013], which learns three dictionaries of different lifespan to tackle the common drift problem in tracking.

3.5 STATISTICAL DICTIONARY LEARNING

The dictionary learning algorithms introduced in the previous sections are mostly formulated as some optimization problems. Besides typical parameters that need to be set for optimization, e.g., dictionary size and sparsity, often, some regularization terms are introduced to the basic formulation in order to enforce some desired properties for the learned dictionary. It is not easy to decouple the impact of the parameters and regularization terms on the final learned dictionary, as the learning process is in general controlled by all these factors. Hence, it is also difficult to predict the generalized performance of the learned dictionary when it is applied to novel problems. To tackle this problem, some efforts have been devoted to formulating the dictionary learning

Algorithm 18 The tracking algorithm via online dictionary learning proposed in Bai and Li [2012]

Input : the video stream $p(x)$, initial dictionary $\mathbf{D}_0$
Output : dictionary $\mathbf{D}$ and sequences of the estimation of the target

1: **for** $t = 1, \cdots, T$ **do**
2:    Sample windows from the new frame according to certain distribute, e.g., particle filter
3:    Apply sparse coding on the sampled windows and compute the sparse representation of those windows
4:    Reject the windows with large reconstruction error or violating sparsity constraint
5:    Use the remaining windows to estimate the target in the new frame
6:    Update the dictionary $\mathbf{D}$ and the distribution with the estimated target
7: **end for**

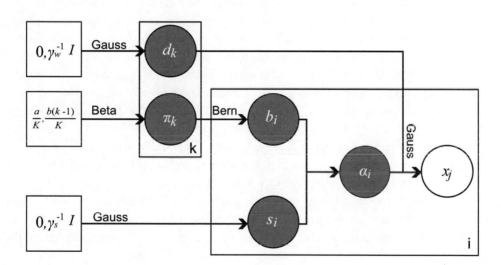

Figure 3.7: The statistic model proposed by Zhou et al. [2009], where gray circle indicates hidden variables, white circle indicates observation and whilte box indicates parameter for the priors.

problem under a statistic learning framework. The basic idea is to replace the hard parameters and regularization terms by priors or distributions, so as to support more flexibility for generalization. These statistical methods share some similarities with the topic models in statistical learning and document analysis. This subsection presents a few such approaches.

One early attempt can be traced to Zhou et al. [2009], where the following statistical model for dictionary learning was proposed, which is also shown in Fig. 3.7:

$$
\begin{aligned}
\mathbf{x}_i &= \mathbf{D}\alpha_i + \epsilon_i & (3.122)\\
\mathbf{D} &= [\mathbf{d}_1, \mathbf{d}_2, \cdots, \mathbf{d}_K] & (3.123)\\
\alpha_i &= \mathbf{z}_i \circ \mathbf{w}_i & (3.124)\\
\epsilon_i &\sim \mathcal{N}(0, \gamma_\epsilon^{-1}\mathbf{I}) & (3.125)\\
\mathbf{d}_k &\sim \mathcal{N}(0, n^{-1}\mathbf{I}) & (3.126)\\
\mathbf{w}_i &\sim \mathcal{N}(0, \gamma_w^{-1}\mathbf{I}) & (3.127)\\
\mathbf{z}_i &\sim \prod_{k=1}^{K} \mathrm{Bern}(\pi_k) & (3.128)\\
\pi_k &\sim \mathrm{Beta}(\frac{a}{K}, \frac{b(K-1)}{K}), & (3.129)
\end{aligned}
$$

where $\mathbf{x}_i \in \mathbb{R}^{n \times 1}$ is the input signal, $\mathbf{D}$ the dictionary with $\mathbf{d}_k$ as its k_{th} atom and α_i the sparse coefficient for signal $\mathbf{x}_i$.

In such a model, the input signal $\mathbf{x}_i$ is sparsely represented by coefficient α_i and dictionary $\mathbf{D}$ subject to some zero-mean Gaussian noise ϵ_i. The sparse coefficient is formed by two components, support vector $\mathbf{z}_i$ and magnitude vector $\mathbf{w}_i$, via element-wise multiplication. The binary support vector $\mathbf{z}_i$ is drawn from the Bernouli distribution, which is parameterized by the Beta process $\mathrm{Beta}(\frac{a}{K}, \frac{a(K-1)}{K})$. Both the dictionary atoms and magnitude vector $\mathbf{w}_i$ are drawn from a zero-mean distribution. Obviously, the sparsity and the support (location of the nonzero elements) of the sparse coefficient α_i are controlled by $\mathbf{z}_i$ and the Bernoulli distribution (parameterized by π_k) and the Beta process (parameterized by $(\frac{a}{K}, \frac{a(K-1)}{K})$); the magnitude of the sparse coefficient instead is determined by $\mathbf{w}_i$ and the Gaussian distribution (parameterized by $(0, \gamma_w^{-1}\mathbf{I})$). One benefit of this model is that, the sparsity and magnitude of the sparse coefficient can be controlled separately, which results in strictly sparse coefficient. In contrast, other dictionary learning algorithms, e.g., these using the ℓ_1 norm for the sparsity constraint, do not yield strictly sparse coefficients (most elements of the coefficient are near-zero but not actually zero). In addition, we do not need to specify the sparsity (the number of nonzero elements) of the sparse coefficient anymore and instead we can use (a, b) to control the sparsity.

For inference, variational inference or Gibbs sampling was used.[4] The full likelihood of the model can be written as:

$$
\begin{aligned}
p(\mathbf{X}, \mathbf{D}, \mathbf{Z}, \mathbf{W}, \pi) = &\prod_{i=1}^{N} N(\mathbf{x}_i | \mathbf{D}(\mathbf{z}_i \circ \mathbf{w}_i), \gamma_\epsilon^{-1}\mathbf{I}) N(\mathbf{w}_i | 0, \gamma_w^{-1}\mathbf{I})\\
&\prod_{k=1}^{K} N(\mathbf{d}_k | 0, n^{-1}\mathbf{I}) \mathrm{Beta}(\pi_k | \frac{a}{K}, \frac{a(K-1)}{K})\\
&\prod_i \prod_k \mathrm{Bern}(z_{i,k} | \pi_k). & (3.130)
\end{aligned}
$$

[4]The detailed algorithm and code can be found in http://people.ee.duke.edu/~mz1/Softwares/.

The posterior regarding dictionary atom $\mathbf{d}_k$ can be expressed as:

$$p(\mathbf{d}_k|\cdots) \propto \prod_{i=1}^{N} N(\mathbf{x}_i|\mathbf{D}(\mathbf{z}_i \circ \mathbf{w}_i), \gamma_\epsilon^{-1}\mathbf{I})N(\mathbf{w}_i|0, \gamma_w^{-1}\mathbf{I}) \sim N(\mu_{\mathbf{d}_k}, \Sigma_{\mathbf{d}_k}), \tag{3.131}$$

where the mean and co-variance of the Gaussian distribution can be computed as:

$$\mu_{\mathbf{d}_k} = \gamma_\epsilon \sum_i z_{i,k} w_{i,k} \mathbf{x}_i^{-k} \tag{3.132}$$

$$\Sigma_{\mathbf{d}_k} = (n\mathbf{I} + \gamma_\epsilon \sum_i z_{i,k}^2 w_{i,k}^2)^{-1}, \tag{3.133}$$

with $\mathbf{x}_i^{-k} = \mathbf{y}_i - \mathbf{D}(\mathbf{z}_i \circ \mathbf{w}_i) + \mathbf{d}_k(z_{i,k} \circ w_{i,k})$.

The posterior regarding the binary support vector $\mathbf{z}_i$ can be expressed as:

$$p(z_{i,k}|\cdots) \propto N(\mathbf{x}_i|\mathbf{D}(\mathbf{z}_i \circ \mathbf{w}_i), \gamma_\epsilon^{-1}\mathbf{I})\text{Bern}(z_{i,k}|\pi_k). \tag{3.134}$$

Thus, $z_{i,k}$ can be drawn from the Bernoulli distribution, with $p(z_{i,k} = 0) = 1 - \pi_k$ and $p(z_{i,k} = 1) = \pi_k e^{\frac{\gamma_\epsilon}{2}(w_{i,k}^2 \mathbf{d}_k^T \mathbf{d}_k - 2w_{i,k}\mathbf{d}_k^T \mathbf{x}_i^{-k})}$.

The posterior regarding the magnitude vector $\mathbf{w}_i$ can be expressed as:

$$p(w_{i,k}|\cdots) \propto N(\mathbf{x}_i|\mathbf{D}(\mathbf{z}_i \circ \mathbf{w}_i), \gamma_\epsilon^{-1}\mathbf{I})N(w_{i,k}|0, \gamma_w^{-1}\mathbf{I}) \sim N(\mu_{s_{i,k}}, \Sigma_{s_{i,k}}), \tag{3.135}$$

where $\Sigma_{s_{i,k}} = (\gamma_w + \gamma_\epsilon z_{i,k}^2 \mathbf{d}_k^T \mathbf{d}_k)^{-1}$ and $\mu_{s_{i,k}} = \gamma_\epsilon \Sigma_{s_{i,k}} z_{i,k} \mathbf{d}_k^T \mathbf{x}_i^{-k}$.

The posterior regarding π_k can be expressed as:

$$p(\pi_k|\cdots) \sim \text{Beta}(\frac{a}{K} + \sum_i z_{i,k}, \frac{bK-1}{K} + N - \sum_i z_{i,k}). \tag{3.136}$$

The whole inference algorithm with Gibbs sampler is described in Algorithm 19. This idea can also be found in Li et al. [2011b]. It was also studied under the online learning framework in Li et al. [2012], where the hyper-parameters γ_w, γ_s, a, and b are sequentially updated via a weighted average between their previous states and the hyper-parameters estimated only from the current mini-batch.

A similar idea was studied in Dobigeon and Tourneret [2010], but with different priors: the dictionary is required to be orthogonal and the sparse coefficient follows the Bernoulli-Gaussian distribution. The prior distribution of the dictionary can be written as follows:

$$p(\mathbf{D}) = \frac{\delta(\mathbf{D} \in \mathcal{S}_{M \times N})}{\text{VOL}(\mathcal{S}_{M \times N})}, \tag{3.137}$$

where $\mathcal{S}$ is the set of the orthogonal matrices in $\mathbb{R}^{M \times N}$ and $\text{VOL}(\mathcal{S}) = \frac{2^M \pi^{\frac{NM}{2}}}{\pi^{\frac{M(M-1)}{4}} \prod_m \Gamma(\frac{N+1-m}{2})}$. The Bernoulli-Gaussian distribution, which encourages sparsity for the sparse coefficient, can

---

Algorithm 19 The inference algorithm for nonparametric dictionary learning proposed in Zhou et al. [2009]

Input : the training data $\mathbf{X}$, parameter a, b, n, γ_w and $gamma_\epsilon$
Output : dictionary $\mathbf{D}$ and sparse coefficient $\alpha = \mathbf{Z} \circ \mathbf{W}$

1: **while** NOT converged **do**
2:   Draw K dictionary atoms $\mathbf{d}_k$ with Eq. 3.131 to form the dictionary
3:   Draw the binary support vector $\mathbf{z}_i$ and magnitude vector $\mathbf{w}_i$ for each training data $\mathbf{x}_i$ with Eq. 3.134 and 3.135 accordingly;
4:   Draw π_k for each k with Eq. 3.136
5:   Check the convergence by computing the full likelihood of the model
6: **end while**

---

be viewed as a mixture of the Bernoulli distribution and the Gaussian distribution: $BG(x) = (1 - \lambda)\delta(x) + \lambda N(x|0, \sigma^2)$. The model is illustrated in Fig. 1 of Dobigeon and Tourneret [2010]. For inference, a Markov Chain Monte Carlo (MCMC) method with Gibbs sampler was proposed, which is similar to Algorithm 19. Interested readers are suggested to check Dobigeon and Tourneret [2010] for more details. Examples of image denoising with the learned dictionary can be found in Fig. 9 of Dobigeon and Tourneret [2010], with comparison to K-SVD.

In Yang et al. [2010b], dictionary learning was combined with the K-mixture model for representing image local descriptors. Consider an image represented by a set of observation points with local descriptor $\mathbf{X} = \{\mathbf{x}_i\}$; it can be modeled as follows:

$$p(\mathbf{X}|\theta) = \prod_i p(\mathbf{x}_i|\theta) = \prod_i \sum_{z_i=1} w_{z_i} p(\mathbf{x}_i|\mathbf{D}_{z_i}), \tag{3.138}$$

where $z_i \in \{1, 2, \cdots, K\}$ is the random variable indicating the mixture assignment, $\theta = \{\mathbf{w}, \mathbf{B}\}$ is the mixture model parameter with $\mathbf{w}$ the mixture weight. The sparse representation is used as the mixture model, accordingly,

$$p(\mathbf{x}_i|\mathbf{B}_j) = \int p(\mathbf{x}_i|\mathbf{B}_j, \alpha_i^j) p(\alpha_i^j|\sigma) d_{\alpha_i^j} = \int N(\mathbf{x}_i - \mathbf{B}_j \alpha_i^j|0, \sigma_1^2\mathbf{I}) L(\alpha_i^j|\sigma) d_{\alpha_i^j}, \tag{3.139}$$

where $\mathbf{B}_j = \mathbf{D}_{z_i}$ $L(\alpha|\sigma) = e^{-\frac{\|\alpha\|_1}{\sigma}}$ is the Laplacian prior. By combing dictionary learning with the mixture model, the method not only produces high-dimensional sparse representations efficiently, but also encourages similar data to take similar sparse representations (the K-mixture model yields a clustering structure for the data). A fast variational inference algorithm was used to learn the dictionary from the training data and find the optimal representation for the data.

Statistic learning was also studies for discriminative dictionary learning in Lian et al. [2010], where the model (also illustrated in Fig. 2 of Lian et al. [2010]) can be interpreted as the following generative process.

1. Draw the probability vector of word assignments θ from Dirichlet distribution $\text{Dir}(\alpha)$.

2. For each image descriptor $\mathbf{w}_i$:

   (a) draw a word assignment z_i from multi-nomial distribution $\text{Mult}(\theta)$ and

   (b) generate a descriptor $\mathbf{w}_i$ from Gaussian distribution $N(\mu_{z_i}, \Sigma_{z_i})$.

3. Draw class label c from distribution $p(c|\bar{\mathbf{z}}, \eta) = \frac{e^{\eta_c^T \bar{z}}}{\sum_l e^{\eta_l^T \bar{z}}})$ with $\bar{z} = \frac{1}{N} \sum_{i=1}^N z_i$ being the empirical distribution of the word assignments.

Accordingly, $\mathbf{D} = [(\mu_i, \Sigma_i)]$ is the dictionary, word assignment $\mathbf{z}$ is the sparse coefficient or feature vector and η_c is the classifier for Class c. The model combines the generative model (Gaussian mixture model) with discriminative model (logistic regression) under a statistic learning framework. A variational inference algorithm was used to learn the proposed model.

The problem of learning non-negative dictionary has also been studied under an statistic learning framework in Dikmen and Févotte [2012]. Consider the data $\mathbf{X} \in \mathbb{R}^{M \times N}$; it can be generated from the following process:

$$x_i^j = \sum_k y_{i,j}^k \tag{3.140}$$

$$y_{i,j}^k \sim \text{Poisson}(w_i^k h_k^j) \tag{3.141}$$
$$\mathbf{w}_k = \text{Dir}(\gamma_k) \tag{3.142}$$
$$h_k^j \sim \text{Gamma}(\alpha_k, \beta_k), \tag{3.143}$$

where $\mathbf{w}_i$ is the coefficient for i_{th} data and $\mathbf{h}_k$ is the k_{th} dictionary atom. By using the fact that the gamma distribution is the conjugate of the Poisson distribution, the posterior of $y_{i,j}^k$ follows a multinomial distribution. Accordingly, a maximal likelihood estimation (MLE) based inference algorithm was used.

In Yang and Yang [2012], discriminative dictionary learning was combined with conditional random field (CRF) for visual saliency detection, where the sparse coefficients were used as latent variables of CRF. The energy of the CRF can be written as:

$$\begin{aligned} E(\alpha, \mathbf{y}, \mathbf{w}, \mathbf{D}) &= \sum_{i \in \mathcal{V}} \psi(\alpha_i, y_i, \mathbf{w}_1) + \sum_{(i,j) \in \mathcal{E}} \phi(y_i, y_j, \mathbf{w}_2) \\ &= \sum_{i \in \mathcal{V}} -y_i \mathbf{w}_1^T \alpha_i + \sum_{(i,j) \in \mathcal{E}} \mathbf{w}_2(i, j) \delta(y_i - y_j), \end{aligned} \tag{3.144}$$

where α is the sparse coefficient, $\mathbf{D}$ the dictionary, $\mathbf{y}$ the pixel label (salient or non-salient) and $\mathbf{w} = \{\mathbf{w}_1, \mathbf{w}_2\}$ the parameter of the CRF. For inference, a max-margin approach was proposed, which alternately updates the dictionary, sparse coefficient and CRF parameters.

The popular robust PCA method [Wright et al., 2009a] was formulated under a Bayesian framework in Ding et al. [2011]. Robust PCA seeks a decomposition of an input matrix $\mathbf{X}$ into a low-rank term $\mathbf{A}$ and a sparse term $\mathbf{E}$:

$$\mathbf{A}, \mathbf{E} : \min_{\mathbf{A}, \mathbf{E}} \|\mathbf{A}\|_* + \lambda \|\mathbf{E}\|_1 \text{ s.t. } \|\mathbf{X} - \mathbf{A} - \mathbf{E}\|_F^2 \leq \epsilon, \qquad (3.145)$$

where $\|\mathbf{X}\|_* = \sum_i \sigma_i(\mathbf{X})$ is the nuclear norm, which is the convex surrogate of the rank of the matrix $\mathbf{X}$, and $\sigma_i(\mathbf{X})$ is the i_{th} singular value of the matrix. Robust PCA stems from the problem where we have an input matrix with limited number of factors (i.e., the matrix has a low rank) with additional sparse noise. In Ding et al. [2011], the input matrix $\mathbf{X}$ is generated from the following process:

$$\mathbf{X} = \mathbf{D}\Delta(\mathbf{z} \circ \lambda)\mathbf{W} + \mathbf{B} \circ \mathbf{Y} + \mathbf{E} \qquad (3.146)$$

$$e_i^j \sim N(0, \gamma^{-1}) \qquad (3.147)$$

$$\mathbf{d}_k \sim N(0, \frac{1}{N}\mathbf{I}) \qquad (3.148)$$

$$\mathbf{w}_n \sim N(0, \frac{1}{K}\mathbf{I}) \qquad (3.149)$$

$$z_k \sim \text{Bernoulli}(p_k) \qquad (3.150)$$

$$p_k \sim \text{Beta}(\alpha_0, \beta_0) \qquad (3.151)$$

$$\lambda_k \sim N(0, \tau^{-1}) \qquad (3.152)$$

$$\mathbf{b}_m \sim \prod_n \text{Bernoulli}(\pi_n) \qquad (3.153)$$

$$\pi_n \sim \text{Beta}(\alpha_1, \beta_1) \qquad (3.154)$$

$$\mathbf{y}_m \sim N(0, \nu^{-1}\mathbf{I}). \qquad (3.155)$$

In this process, $\mathbf{B}$ is the binary matrix controlling the sparsity of the noise, $\mathbf{z}$ is the binary vector controlling the rank of the matrix and $\mathbf{z}$ controlling the "magnitude" of the singular values of the matrix. By assuming $\mathbf{d}_k \sim N(0, \frac{1}{N}\mathbf{I})$, we have $\text{var}(\mathbf{d}_k) = \mathbf{I}$ or equivalently $\|\mathbf{d}_k\|_2 = 1$. Thus, $\mathbf{D}\text{Diag}(\mathbf{z} \circ \lambda)\mathbf{W}$ simulates a singular value decomposition. Compared with robust PCA, this statistical model is able to generate the matrix which is a mixture of a truly low-rank term and a truly sparse term. For inference, MCMC with Gibbs sampler and variational Bayesian inference were proposed and compared. Interested readers should refer to Sec. IV of Ding et al. [2011] for the details. For video applications, where each frame is treated as a column of the input matrix, Markovian dependency was further introduced from the sparsity term $\mathbf{B}$ in both space and time. Examples of applying robust PCA in background subtraction for video, where the background is extracted in the low-rank term and foreground is extracted in the sparse term, are shown in Fig. 3.8.

Coupled dictionary learning via a statistic learning framework was studied in He et al. [2013]. Compared with other coupled dictionary learning methods, e.g., Yang et al. [2008,

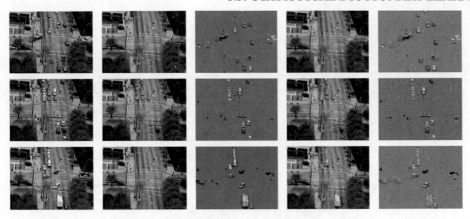

Figure 3.8: Examples of applying robust PCA in background subtraction for video (frames shown in Column 1), where the background is extracted in the low-rank term (Columns 2 and 4) and the foreground is extracted in the sparse term (Columns 3 and 5). For comparison, the results of Ding et al. [2011] are shown in Column 2 and 3; the results of Wright et al. [2009a] are shown in Columns 4 and 5. Three examples are shown in three rows accordingly. This figure is from Fig. 10 of Ding et al. [2011].

2010a], which typically require the sparse coefficient of the two coupled spaces to be identical, a more flexible constraint is introduced, where the sparse coefficients are not required to be identical but to share the same sparse support. The sparse support is drawn from a Beta process and the values of the sparse coefficients are drawn from two different Gaussian priors. Accordingly, the data from two coupled spaces $\mathbf{x}_i$ and $\mathbf{y}_i$ can be generated from the following process:

$$\mathbf{x}_i = \mathbf{D}^x \alpha_i^x + \epsilon_i^x \quad , \quad \mathbf{y}_i = \mathbf{D}_i^y \alpha_i^x + \epsilon_i^y \tag{3.156}$$

$$\alpha_i^x = \mathbf{z}_i \circ \mathbf{s}_i^x \quad , \quad \alpha_i^y = \mathbf{z}_i \circ \mathbf{s}_i^y \tag{3.157}$$

$$\mathbf{d}_k^x \sim N(0, P_x^{-1}\mathbf{I}) \quad , \quad \mathbf{d}_k^y \sim N(0, P_y^{-1}\mathbf{I}) \tag{3.158}$$

$$\mathbf{s}_i \sim N(0, \gamma_{s,x}^{-1}\mathbf{I}) \quad , \quad \mathbf{s}_j \sim N(0, \gamma_{s,y}^{-1}\mathbf{I}) \tag{3.159}$$

$$\mathbf{z}_i \sim \text{Bernoulli}(\pi_k) \quad , \quad \pi_k \sim \text{Beta}(\frac{a}{K}, \frac{b(K-1)}{K}) \tag{3.160}$$

$$\epsilon_i^x \sim N(0, \gamma_{\epsilon,x}^{-1}\mathbf{I}) \quad , \quad \epsilon_i^y \sim N(0, \gamma_{\epsilon,y}^{-1}\mathbf{I}) \tag{3.161}$$

$$\gamma_{s,x}, \gamma_{s,y} \sim \text{Gamma}(c, d) \quad , \quad \gamma_{\epsilon,x}, \gamma_{\epsilon,y} \sim \text{Gamma}(e, f), \tag{3.162}$$

where $\mathbf{z}_i$ is the sparsity support shared by the sparse coefficients for the two coupled spaces. This model is also visualized in Fig. 3.9. The experimental result of single image super-resolution with the dictionary learned with He et al. [2013] with comparisons to several state-of-art methods can be found in Fig. 3 of He et al. [2013].

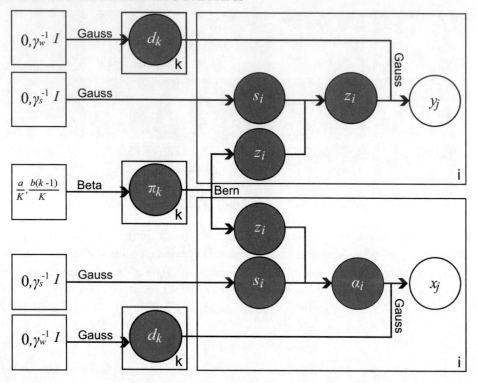

Figure 3.9: The model proposed in He et al. [2013], where the sparse coefficient of the two feature spaces share the same sparsity support $\mathbf{z}_i$.

To model the relationship between the image and rich text descriptions associated with the image, Irie et al. [2013] proposed Continuous-Discrete Bayesian Co-clustering (CD-BCC). CD-BCC finds out the clustering of the visual descriptor and their relationship to the text descriptor via a relational matrix. An illustration of the model can be found in Fig. 2 of Irie et al. [2013], where for the visual descriptor, a generative process similar to Zhou et al. [2009] is used. For the relational matrix, the Poisson process is used.

1. Draw the co-occurrence between visual descriptor and text $\theta_{k,l} \sim \text{Gamma}(\beta, \phi)$.

2. Draw visual descriptor cluster distribution $k \sim \text{Dir}(\frac{\zeta}{K})$ and text cluster distribution $\lambda \sim \text{Dir}(\frac{\eta}{K})$.

3. For each descriptor distribution v and text j, draw image descriptor cluster assignment $z_v^x \sim \text{Multi}(k)$ and text cluster assignment $z_j^w \sim \text{Multi}(\lambda)$.

4. For each pair of visual descriptor and word distribution (v, j), draw element of the relational matrix as $r_{v,j} \sim \text{Poisson}(\theta_{z_v^x, z_j^w})$.

For inference, MCMC and Gibbs sampler are used. By combining visual features of an image and the associated text descriptor, CD-BCC is able to learn a more discriminative descriptor.

While most statistic learning methods described above assume that the inputs are drawn independently, efforts have also been attempted to model the dependency of the data. In Zhou et al. [2011], nonlocal statistics of the data are modeled via a dependent hierarchical Beta process, where similar patches are required to take similar sparse coefficients over the dictionary. To this end, the covariance of the sparse coefficients are modeled. The hierarchical model which generated the data can be written as:

$$\mathbf{x}_i \sim N(\mathbf{D}(\mathbf{s}_i \circ \mathbf{z}_i), \gamma_\epsilon^{-1}\mathbf{I}) \tag{3.163}$$

$$\mathbf{d}_k \sim N(0, P^{-1}\mathbf{I}) \tag{3.164}$$

$$\mathbf{s}_i \sim N(0, \gamma_s^{-1}\mathbf{I}) \tag{3.165}$$

$$z_{i,k} \sim \text{Bernoulli}(\pi_{i,k}) \tag{3.166}$$

$$\pi_{i,k} = \sum_{j \in \mathcal{Q}_i} a_{i,j} \pi_{j,k}^* \tag{3.167}$$

$$\pi_{j,k}^* \sim \text{Beta}(c_1 \eta_k, c_1(1 - \eta_k)) \tag{3.168}$$

$$\eta_k \sim \text{Beta}(c_0 \eta_0, c_0(1 - \eta_0)), \tag{3.169}$$

where $\mathcal{Q}_i$ defines a set of indices in the (nonlocal) neighborhood of i and, accordingly, $\pi_{i,k} = \sum_{j \in \mathcal{Q}_i} a_{i,j} \pi_{j,k}^*$ computes the weighed average of the nonlocal neighbors. For inference, MCMC was used. Examples of image denoising with dictionary learned by Zhou et al. [2009, 2011] are shown in Fig. 3 of Irie et al. [2013]. This model was further studied in Zhou et al. [2012a], where the nonlocal statistics were modeled via hierarchical Dirichlet process. In Peleg et al. [2012], the dependency of the data is modeled via Boltzmann machine, which is inferred via a message passing algorithm.

CHAPTER 4

Applications of Dictionary Learning in Visual Computing

The general idea of sparse representation based on dictionaries learned adaptively from data has seen numerous applications. In many cases, the current literature suggests that significant progresses have been obtained by recent approaches based on the general dictionary learning idea, when compared with more conventional approaches. As the focus of this book is on visual computing applications, we now illustrate how the general idea has been adapted to different visual computing tasks. Along the way, by summarizing some of the key technological components of the approaches reviewed below, we also intend to bring a reader's attention to those application-specific techniques or sometimes seemingly small "tweaks" that are often essential to the successful deployment of the general idea.

We cover the following general topics in visual computing: image compression, reconstruction of degraded images, image super-resolution, segmentation in images, image-based classification, visual saliency detection, and visual tracking. In a couple of cases (e.g., compression), we also go beyond the visual domain to briefly discuss other modalities, due to the close relationship between the underlying methodologies and, accordingly, the benefit of making such relationship visible by including such discussion in the context.

4.1 SIGNAL COMPRESSION

In this section, we introduce several efforts on applying dictionary learning for signal compression, including general image compression (Sec. 4.1.1), face image compression (Sec. 4.1.2), and audio signal compression (Sec. 4.1.3).

4.1.1 IMAGE COMPRESSION

One of the first applications of dictionary learning is image compression [Aharon et al., 2005]. Images may be considered as a collection of patches/blocks, as is done in most existing image codec systems (a review of using vector quantization in image compression can be found in Nasrabadi and King [1988]). At the patch level, images are repetitive [Zontak and Irani, 2011]. To capture this redundancy, many methods represent an image patch as a linear combination of a few dictionary atoms, where the dictionary is learned from many training patches. Then the image is represented by the sparse coefficients of all the patches. Sparse coefficient can be stored by only

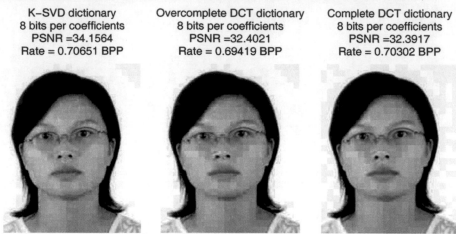

Figure 4.1: The example of compression with dictionary learned by K-SVD (left), overcomplete DCT dictionary (middle) and complete DCT dictionary (i.e., JPEG, right). In this example, the patches are 8 × 8 blocks extracted from the images without overlap (thus boundary of the patches can be observed from the results). The dictionary consists of 441 atoms learned from 594 patches. This figure is from Fig. 9 of Aharon et al. [2005].

keeping the locations and the values of its nonzero elements. To further improve the compression ratio, encoding methods can be used for store the locations and values of the sparse coefficient, e.g., by entropy coding. To reconstruct the image from the compressed data, each patch is reconstructed by using the sparse coefficient. An example of compression results with K-SVD with comparison to JPEG, where DCT dictionary is used, is shown in Fig. 4.1, where the results with K-SVD shows an improvement of 1–2 db over DCT dictionary, when the same compression ratio is used. In such example, the patches are extracted without overlapping, which leads to the block artifact as shown in the results. To reduce the block artifact, patches may be extracted from the image with overlap and the overlapped region is reconstructed by computing the mean of the reconstructed patches.

The compression ratio measures the ratio of the size of compressed data over the size of the input image. For dictionary learning-based image compression, not only the sparse coefficient but also the learned dictionary is required to represent the compressed image. Thus, the compression ratio (bit per pixel, or bpp) can be computed as:

$$r = \frac{\frac{MN}{B^2}k(a+b) + B^2Tc}{dMN} = \frac{k}{B^2}\frac{a+b}{d} + \frac{B^2T}{MN}\frac{c}{d}, \qquad (4.1)$$

where the input image is $M \times N$, a patch is $B \times B$, k is the (average) number of nonzero elements for the sparse coefficient, T is the size of the dictionary, a, b, c, and d are, respectively, the number of bits required to store the indexes of non-zero element in the sparse coefficient, the number of bits required to store the value of the nonzero-element, the number of bits required to store the value of the dictionary atom, and the number of bits required to store the value of one pixel. $B = 8$ is the most common choice for the patch size. From this equation, we can find that to increase the compression ratio, we need to reduce k (the average number of atoms used for reconstructing a patch) and T (the number of atoms in the dictionary). However, k also determines the quality of the reconstruction: a smaller k results in higher compression ratio but lower reconstruction quality. Consider the following example $B = 8$, $T = 256$, and $a, b, c, d = 8$ (gray image), we have $r = \frac{k}{32} + \frac{128^2}{MN}$. When the image is smaller than 128×128, the compression ratio cannot be smaller than 1, and for comparison the compression ratio for JPEG with 25% quality is typically 1. As a result, this method is not useful for small images.

Another approach was proposed in Dobigeon and Tourneret [2010], where the dictionary atoms are required to be orthogonal and learned via Bayesian inference. Experimental results on the Barbara image show improved image quality (smaller root of mean squared error) with the same compression ratio, compared with K-SVD, when 16×16 patches are used.

The performance of the compression relies on the goodness of the learned dictionary. The recursive least square dictionary learning algorithm (RLS-DLA) [Skretting and Engan, 2011] learns a dictionary from a large set of training patches via online learning. By using a forget parameter, RLS-DLA is not sensitive to a bad initial dictionary, which however is an issue for many other dictionary learning algorithms. This results in improved PSNR over K-SVD, as shown in Table 4.1.

Table 4.1: Comparison of PSNR for several image compression algorithms on Lena image with varying compression ratio. For MOD (wavelet), K-SVD (wavelet) and RLS-DLA (wavelet) wavelet multiscale decomposition is used. This table is from Table 1 of Skretting and Engan [2011].

Compression ratio (bpp)	0.25	0.5	0.75	1	1.5
MOD (patch)	34.25	37.51	39.42	40.81	42.93
K-SVD (patch)	34.19	37.52	39.48	40.92	43.14
RLS-DLA (patch)	34.22	37.76	39.82	41.32	43.65
DCT	34.02	37.41	39.54	41.09	43.44
MOD (wavelet)	35.12	38.21	39.98	41.28	43.34
K-SVD (wavelet)	35.13	38.26	40.07	41.41	43.56
RLS-DLA (wavelet)	35.26	38.57	40.42	41.78	44.02
9/7 wavelet	35.08	38.29	40.19	41.61	43.9
JPEG	31.64	35.85	37.77	39.14	41.16
SPIHT	34.78	38.11	40.12	41.4	43.26
JPEG-2000	35.29	38.6	40.48	41.91	44.13

Multi-scale dictionaries were used to improve the reconstruction quality in Ophir et al. [2011]. For compression, wavelet transform is applied to the image to build a pyramid, where each level includes four bands LL, LH, HL, HH (L for low frequency and H for high frequency); the patches are extracted from the LH, HL, HH bands of each level without overlapping; and each patch is then sparse-coded, resulting in the final representation of the image. Examples of compression results on fingerprint images and natural images are shown in Figs. 7 and 8 of Ophir et al. [2011], where the following settings are used: three levels of pyramid, which results in 10 dictionary (3 for each level and the extra one for the top level), 8×8 patches and 64 atoms for each dictionary, and sparsity constraint is 5. Experiments show obvious improvement on image quality (a gain of 4–6 dB). A similar idea was used in Mazaheri et al. [2013], where a hierarchical dictionary was learned and used for image compression. However, only one atom can be chosen for each level of the dictionary.

In Jiang et al. [2012], the compression performance of different types of dictionaries, e.g., DCT, wavelet, Gabor and dictionary learned adaptively from data, are compared. The dictionaries were built by greedily selecting elements from a set of predefined atoms, which could be DCT, Haar wavelet, Daub 4 wavelet, Gabor and image patches. Experiments on image compression (Fig. 3c of Jiang et al. [2012]) show that the the dictionary learned adaptively from the data achieves the best results, followed by DCT, wavelet, and Gabor, when the same sparsity constraint is used.

4.1.2   FACE IMAGE COMPRESSION

While the compression algorithms for natural image can also be applied to face images, one may exploit the special properties of face images to drive up the compression ratio while maintaining high image quality. In Bryt and Elad [2008], face images are aligned to some canonical form and patches of size 15×15 are extracted from face images at fixed locations without overlap (for an visual illustration, the readers could refer to Fig. 2 of Bryt and Elad [2008]). The patches of each location are then compressed with their unique dictionary. Due to the redundancy of the face images, the dictionaries can be shared for all face images and the patches can be represented with much sparser coefficient.

The dictionary for each location contains 512 atoms (a sample dictionary for patches at left eye can be found in Fig. 6 of Bryt and Elad [2008]) and each patch is represented by only 4 atoms. Accordingly, the compression ratio is $r = \frac{2 \times 4}{15^2} = 0.035$, where the storage of dictionaries can be negligible as they can be shared for all face images. The sparse coefficient is further coded by a straightforward Huffman table to improve the compression ratio. It was reported in Bryt and Elad [2008] that offline training takes 10 h, compression of a 179×221 image takes 5 s and decompression takes 1 s. The experimental results with comparison to other methods are shown in Fig. 4.2, where the method of Bryt and Elad [2008] achieved impressive results.

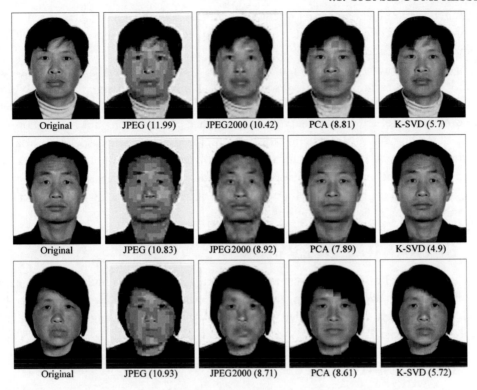

Figure 4.2: Comparison of different compression algorithm on face images. For each row, from left to right: input image, compression results with JPEG, JPEG 2000, PCA, and the method proposed in Bryt and Elad [2008]. The face image is compressed to 820 bytes and the number in brackets is RMSE. This figure is from Fig. 9 of Aharon et al. [2006].

4.1.3 AUDIO SIGNAL COMPRESSION

Dictionary learning has also been used for compressing audio signal [Yaghoobi et al., 2009a,b]. Similarly, the audio signal is divided into blocks and each block has 1024 samples and each block is sparsely coded. Considering the limited frequency range of the human auditorial system, the audio signal is first down-sampled to 32 KHz before sparse coding. The dictionary consists of 2048 atoms and each block is sparsely represented by 40 atoms. The location and the magnitude of the non-zero coefficients are stored. For training, 8192 blocks are used. A comparison of distortion rate of different methods with different compression rate can be found in Fig. 4 of Yaghoobi et al. [2009a], and the examples of the dictionary learned from audio signal blocks can be also found in Fig. 1 of Jafari and Plumbley [2011].

While most dictionary learning algorithms learn nonparametric dictionaries, it has been found in simulation experiments that learning dictionary atoms with structures similar to the

Gabor functions results in better representation of noisy speech signals [Ataee et al., 2010] than the nonparametric dictionaries. Such a dictionary can be compactly represented by a few parameters for the Gabor functions.

Instead of dividing the audio signal into blocks and compressing (sparse coding) each block independently, it was proposed in Mysore [2012] to model audio signal by combining sparse coding and Hidden Markov Models (HMMs). Each block of an audio signal is represented as a linear combination of a few dictionary atoms via sparse coding. Considering the non-stationary property of the signal, multiple dictionaries are employed, where the association of a dictionary and some audio frame is modeled via an HMM as the state and the observation respectively. The sparse coefficients of frames corresponding to each dictionary naturally form block sparsity. Sample dictionaries learned can be found in Fig. 4 of Mysore [2012], where 18 dictionaries are shown and each dictionary has 10 atoms as the rows.

4.2   SIGNAL RECOVERY

Dictionary learning has also been used in many signal recovery tasks including image denoising, image inpainting, image demosaicing, which will be introduced in Sec. 4.2.1, Sec. 4.2.2, and Sec. 4.2.3, respectively. An earlier review on dictionary learning in image processing can also be found in Elad et al. [2010].

4.2.1   IMAGE DENOISING

Image denoising aims to remove noise from an image. The location of the noise is in general unknown. Image noise may come from many sources, such as sensor noise, transmission error, encoding artifacts, etc., resulting in different noise types. Figure 4.3 shows the Lena image under Gaussian noise, salt-and-pepper noise and JPEG noise/artifacts, respectively. Gaussian noise is a widely-used type that has been assumed by many algorithms, although some algorithms also explicitly deal with other types of noise.

Mathematically, the denoising problem can be formulated as:

$$\mathbf{x} : \min_{\mathbf{x}} \|\mathbf{y} - \mathbf{x}\|_p + \lambda \psi(\mathbf{x}), \tag{4.2}$$

where $\mathbf{x}$ is the denoised signal, $\mathbf{y}$ is the noisy signal and $\psi(\mathbf{x})$ is the prior on the denoised signal. For example, by assuming that the signal is mostly smooth except around sparse discontinuities, total variation (TV) can be used for the prior, where $\text{tv}(\mathbf{x}) = \sum_i \|\mathbf{x}_i - \mathbf{x}_{i-1}\|$. However, choosing a good prior is not an easy task and many ideas have been proposed to this end, some of which are introduced below.

For reconstruction error measurement, the ℓ_p norm is often used, where $\ell_p(x) = (\sum_i \|x\|^p)^{\frac{1}{p}}$. p can be 0, 1, or 2, whose choice depends on the noise model: $p = 2$, when Gaussian noise is assumed, is the most common choice [Bao et al., 2013a, Hawe et al., 2013, Rubinstein et al., 2010, Shi et al., 2011, Yaghoobi et al., 2012]; $p = 0$ or $p = 1$ makes the algorithm robust

(a)   (b)   (c)

Figure 4.3: The example of Lena image under different types of noise: Gaussian noise (a), salt-and-pepper noise (b), and JPEG codec artifacts (c).

over some non-Gaussian noise, e.g., salt-pepper noise or impulse noise [Dong et al., 2011a, Liu et al., 2013a, Wang et al., 2013b]. In Dong et al. [2011a], experiments on the Lena image with mixed Gaussian and impulse noise show that the results with $p = 1$ have significant improvement in PSNR over the results with $p = 2$.

Most dictionary learning-based image denoising algorithms rely on image patches and make the assumption that the patches can be sparsely represented by certain dictionary, i.e.,

$$\mathbf{x}, \alpha \min_{\mathbf{x}, \alpha} \|\mathbf{y} - \mathbf{x}\|_p + \lambda \psi(\mathbf{x}) \text{ s.t. } \mathbf{x}_i = \mathbf{D}\alpha_i, \|\alpha_i\|_q < \epsilon, \quad (4.3)$$

where $\mathbf{D}$ is the dictionary which can be learned offline from some training data or even from the noisy image itself, α_i is the sparse coefficient for patch $\mathbf{x}_i$, and q could be 0 or 1 or others for the sparsity constraint, e.g, $(2, 1)$ or $(0, \infty)$. Often, overlapped patches are extracted for processing, in order to reduce blocking artifacts around the patch boundaries. Accordingly, the overlapped regions are updated by the mean of the denoised patches.

Such denoising algorithms share some similarity to the nonlocal means algorithm (NLM) [Mahmoudi and Sapiro, 2005, Peyré, 2011] for image denoising, which computes the value for the denoised pixel as the weighted average of some reference pixels, with weights determined by the similarity between the patch centered at the current pixel and the patch centered at the reference pixel. In a dictionary learning-based image denoising algorithm, each patch is computed as the weighted average (linear combination) of the dictionary atoms, which itself is a patch. The weight or the value of the coefficient is computed via some sparse coding algorithm. Recall that in OMP, this value is computed as the correlation score of the input patch to the dictionary atom, which of course can be viewed as some similarity measure (cosine similarity). However, the weight used in the dictionary learning-based denoising algorithm can be negative. This is in contrast to NLM, in which the weights are always nonnegative.

While some algorithms denoise the patches independently [Aharon and Elad, 2008, Bao et al., 2013a, Charles et al., 2011, Hawe et al., 2013, Liu et al., 2013a, Peng et al., 2013, Rubinstein et al., 2010, Shi et al., 2011, Wang et al., 2013b, Yaghoobi et al., 2012, Yang et al., 2013b], others try to explore the nonlocal statistics of the patches, e.g., adjacent patches should take similar sparse coefficient. In Dong et al. [2011a,b], Feng et al. [2011b], and Li and Fang [2010], the patches are first divided into several groups according to their similarities via some clustering algorithm. Then a dictionary is learned for each group and the patches are sparsely coded with the dictionary of its own group. As a result, similar patches are denoised in a similar manner. Experiments on different types of images (Fig. 4.4) show improvement over BM3D [Dabov et al., 2006] as well as those dictionary-learning-based denoising algorithms without such spatial information, in terms of both PSNR and SSIM [Wang et al., 2004].

In Zhou et al. [2012a, 2009, 2011], image denoising is formulated under a Bayesian inference framework. Considering that the patches of a natural image usually form clusters, Dirichlet-process-based nonparametric clustering is used to cluster the patches into groups. The patches in each group is encouraged to have similar usage of dictionary atoms, i.e., similar sparse coefficients. In addition, patches which are spatially close are also encouraged to take similar sparse coefficient, via the stick breaking process.

The non-local statistics of the patches can also be modeled via the group sparsity constraint [Mairal et al., 2009b]. The $\ell_{p,q}$ norm (defined in Eq. 2.7) with $(p, q) = (2, 1)$ or $(\infty, 0)$ (used in Mairal et al. [2009b]) encourages similar set of atoms to be used for sparse coding for each group, achieving group sparsity. After clustering the patches into different groups, group sparsity coding is applied to find their sparsity coefficient:

$$\alpha : \min_{\alpha} \sum_i \frac{\|\alpha_i\|_{p,q}}{|\mathbf{S}_i|} \text{ s.t. } \sum_{j \in \mathbf{S}_i} \|\mathbf{y}_j - \mathbf{D}\alpha_{i,j}\|_2^2 \leq \epsilon_i \,\forall i, \tag{4.4}$$

where $\mathbf{S}_i$ is Group i, $\alpha_{i,j}$ is the sparse coefficient of Patch j in Group i and $\alpha_i = [\alpha_{i,1}, \alpha_{i,2}, \cdots, \alpha_{i,|\mathbf{S}_i|}]$.

Multi-scale dictionaries have also been studied in image denoising to exploit the multi-scale framework for enhancing the performance. In Yang et al. [2013b], a multi-scale dictionary is learned from the decomposed multi-scale ridgelet support vectors of natural images. The ridgelet, which is an extension of wavelet to higher dimensionality, is better at approximating edges and corners:

$$\psi_t(x) = a^{-\frac{1}{2}}\psi(\frac{u \cdot x - b}{a})(K_{\psi} = \int \frac{\|\hat{\psi}(\epsilon)\|^2}{\|\epsilon\|^d}d\epsilon < \infty), \tag{4.5}$$

where the parameter (a, u, b) defines the scale, orientation and location of the ridgelet, respectively. The learned multi-scale dictionary is then employed for image denoising. The method was reported to be able to preserve better edges, contours, and textures in images while reducing avoiding the ringing effects.

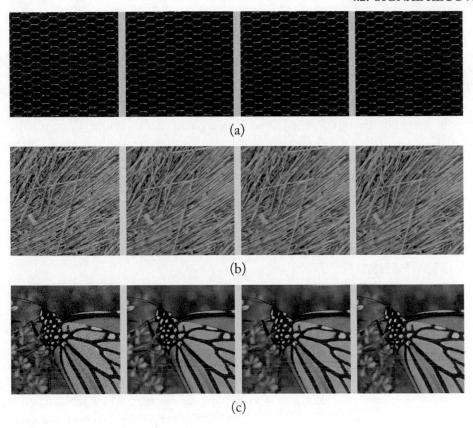

Figure 4.4: Comparison of denoising result on three type of images (a, b, and c) with different algorithms (left to right): input noise images, results of BM3D, results of K-SVD and results of Dong et al. [2011a]. These images are subjected to Gaussian noise with $\sigma = 20$. For numerical measurement, the PSNR is (a) 29.33dB, 30.53dB, 31.30d; (b) 27.09dB, 26.95dB, 27.50dB; (c) 30.37dB, 39.89dB, 30.7dB and the SSIM is (a) 0.9178, 0.9327, 0.9426; (b) 0.8963, 0.8899, 0.9061; (c) 0.9209, 0.9075, 0.9197. These figures are from Figs. 4, 5, and 6 of Dong et al. [2011a].

The idea of multi-scale dictionary was also studied in Aharon and Elad [2008], where the dictionary itself is an "image" and the patch extracted from any location at any scale of that image forms the atom of the dictionary. This dictionary was referred to as image signature dictionary (ISD). Compared with dictionaries learned by other algorithms, the ISD has some benefits including smaller memory footprint and greater flexibility. For sparse coding, a modified matching pursuit algorithm was proposed: the inner product between the input signal and all dictionary atoms is replaced by an efficient operation of convolution between the input signal and the dictionary.

Dictionary learning has also been applied for denoising multi-spectral images [Charles et al., 2011, Peng et al., 2013]. Unlike in the cases of grayscale images or RGB images, 3D blocks instead of 2D patches are extracted from hyper-spectral images and a tensor dictionary is learned accordingly [Peng et al., 2013]. The tensor dictionary is assumed to be decomposable: $\mathcal{D} = \mathcal{Z} \times_1 \mathbf{D}^W \times_2 \mathbf{D}^H \times_3 \mathbf{D}^S$, where $\mathbf{D}^W, \mathbf{D}^H$, and $\mathbf{D}^S$ are the dictionaries for spatial and spectral dimensions, respectively. The correlation among the spectral bands, together with spatial correlation, are exploited to facilitate denoising via group sparse coding. Experiments (Fig. 4 of Peng et al. [2013]) on denoising hyper-spectral images with comparisons to several existing methods, e.g., BM4D (extension of BM3D on hyper-spectral image), show some promising results. In Patel et al. [2014], a dictionary learning based approach was proposed for removing speckle noise in synthetic aperture radar (SAR) image, where the structure and texture component of the image is separately estimated with different dictionaries to sustain the texture information of the image.

4.2.2 IMAGE INPAINTING

Dictionary learning can be applied to recover not only data corrupted by noised but also missing data, making it appropriate for image inpainting, where missing pixels from an image are to be recovered. An example of images with missing pixels is shown in Fig. 4.5 (a) (or c for the zoomed portion), where the pixels covered by the red texts need to be recovered. Similar to dictionary-learning-based image denoising, image inpainting is often done patch-wise, where an image is divided into many patches, the missing pixels in each patch are then reconstructed via sparse representation over some dictionary, and finally the recovered patches are stitched together to reconstruct the entire image. To reduce block artifacts around the patch boundaries, overlapping patches can be extracted. Mathematically, this process can be written as:

$$\mathbf{x}, \alpha \min_{\mathbf{x}, \alpha} f(\mathbf{y} - \mathbf{x}) + \lambda \psi(\mathbf{x}) \text{ s.t. } \mathbf{x}_i = \mathbf{D}\alpha_i, \|\alpha_i\|_q \leq \epsilon, \tag{4.6}$$

where $\mathbf{y}$ is the image with missing pixels, $\mathbf{x}$ the inpainted image, $\psi(\cdot)$ the prior on the inpainted image, $q = \{0, 1\}$, α_i the sparse coefficient for the i_{th} patch $\mathbf{x}_i$ and $\mathbf{D}$ the dictionary. $f(\mathbf{y} - \mathbf{x})$ constrains the inpainted image to be similar to the input image for the "good" pixels. Here the inpainted image is not required to be identical to the input image for the "good" pixels, because other types of noise, e.g., Gaussian noise, may also be present in the input image.

One intuitive choice for f would be $f(\mathbf{y} - \mathbf{x}) = \sum_{(i,j) \in \Omega} \|y_i^j - x_i^j\|_2^2$, where Ω is the set of "good" pixels. However, often we may find such set Ω may not be available. Instead, a sum of squared error may be used $f(\mathbf{y} - \mathbf{x}) = \|\mathbf{y} - \mathbf{x}\|_2^2$ [Bao et al., 2013a, Li et al., 2012, Mairal et al., 2009a, Shi et al., 2011, Zhou et al., 2009]. By sparsely representing the patches over a dictionary whose atoms do not contain missing pixels, the missing pixels in the reconstructed patches could still be recovered—under the assumption that the missing pixels are sparsely distributed in the image, i.e., only a small fraction of the pixels in each patch is missing. Figure 4.5 shows examples of image inpainting with the algorithm from Mairal et al. [2009a].

Figure 4.5: Example of image inpaint with method from Mairal et al. [2009a]: (a) input image, (b) restored image, (c), and (d) a zoomed portion of (a) and (b). These figures are from Fig. 2 of Mairal et al. [2009a].

Compared with the ℓ_2 norm in $f(\mathbf{y} - \mathbf{x}) = \|\mathbf{y} - \mathbf{x}\|_2^2$, a more robust way is to use the ℓ_0 or ℓ_1 norm, such that $f(\mathbf{y} - \mathbf{x}) = \|\mathbf{y} - \mathbf{x}\|_1$ [Lu et al., 2013]. As analyzed in the previous sections, the ℓ_1 or ℓ_0 norms are more appropriate for modeling the sparse noise than the ℓ_2 norm. Missing pixels in image inpainting are typically assumed to be sparse, which can be better captured via the ℓ_1 or ℓ_0 norm. Examples of image inpainting with the ℓ_1 norm can be found in Fig. 3 of Lu et al. [2013]. In Chen and Wu [2013], the error between the input image $\mathbf{y}$ and the recovered image $\mathbf{x}$ is decomposed into two components: $\mathbf{n}$, which captures the Gaussian noise and is regularized by the ℓ_2 norm, and $\mathbf{e}$, which captures the sparse residual for the missing pixels and is regularized by the ℓ_1 norm.

In Szabó et al. [2011], nonlocal statistics were explored to improve the quality of the restoration. Similar to Mairal et al. [2009b], which combines nonlocal statics with dictionary learning for image denoising, the method proposed in Szabó et al. [2011] divides the patches of the input image into groups and uses $\ell_{p,q}$ to enforce the group sparsity for those patches.

A special scenario of image inpainting was studied in Kang et al. [2012], where the input image is captured from a raining scene and the impact of the rain needs to be removed. The problem was formulated as an image decomposition problem. The image is first decomposed into two parts, the low-frequency part and the high-frequency (HF) part, via bilateral filtering. The rain is supposed to exist in the HF part and thus the HF part is further decomposed into a "rain" part and a "non-rain" part via dictionary learning and sparse coding. More specifically, in dictionary learning, patches are extracted from the HF part and a dictionary is learned from those patches. The atoms of the dictionary are clustered into two groups via K-means, where one group of atoms is identified as the "rain" dictionary and the other group is the "non-rain" group, via a histogram-of-gradient (HoG) descriptor. In the sparse coding stage, each patch is sparse-coded on the combined "rain" dictionary and "non-rain" dictionary, then the rain-removed patch is recovered by reconstruction with only the "non-rain" dictionary. The rain-removed patch is combined with the low frequency part to build the final image.

4.2.3 IMAGE DEMOSAICING

Most consumer-grade digital cameras rely on color filters in conjunction with a common imaging array to capture color images. The most widely known color filter uses the Bayer color pattern (Fig. 4.6). The task of recovering three full-resolution RGB color images from a Bayer-filtered raw image is called image demosaicing. Image demosaicing is similar to image inpainting in that they both aims to recover the missing pixels of the images. However, in image demosaicing, the locations of the missing pixels are fixed and known.

The basic idea behind most demosaicing approaches is interpolation [Gunturk et al., 2002, Paliy et al., 2007]. Dictionary learning-based methods have also been proposed for image demosaicing [Mairal et al., 2009b, 2008b]. In Mairal et al. [2008b], a dictionary $\mathbf{D}_0$ is first learned from the patches of some color images in the training set; then the patches of the input image $\mathbf{y}_i$ (or the mosaiced image, where the missing pixels is set to 0) is sparse-coded and reconstructed via the learned dictionary $\hat{\mathbf{x}}_i$:

$$\alpha_i \quad : \quad \min_{\alpha_i} \frac{1}{2}\|\mathbf{M}_i \circ (\mathbf{y}_i - \mathbf{D}_0\alpha_i)\|_2^2 \text{ s.t. } \|\alpha_i\|_0 \leq t \quad (4.7)$$

$$\hat{\mathbf{x}}_i \quad = \quad \mathbf{D}_0\alpha_i, \quad (4.8)$$

where $\mathbf{M}_i$ is a binary mask corresponding to the Bayer pattern for patch $\mathbf{y}_i$, such that only the non-missing pixels are counted. After the initial demosaiced image $\hat{\mathbf{x}}$ is obtained, an image-adaptive dictionary $\mathbf{D}_1$ is learned from its patches. The final demosaiced image is computed with the combined dictionary $\mathbf{D} = [\mathbf{D}_0, \mathbf{D}_1]$ via Eq. 4.7 and 4.8. This problem can be repeated for several iterations to enhance the demosaicing quality.

The above idea was further studied in Mairal et al. [2009b], where image non-local statistics were used to improve the performance. To this end, the patches of the mosaiced image are

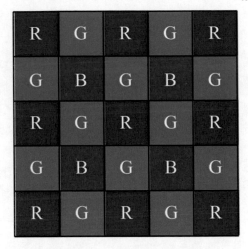

Figure 4.6: The Bayer pattern, where each 2 × 2 block contains two green pixel, one red pixel and one blue pixel.

first clustered into different groups, and then group sparse coding (the $\ell_{\infty,0}$ norm) is applied for patches of each group. By using group sparse coding, similar patches (i.e., patches from the same group) are reconstructed with a set of similar dictionary atoms, which makes the algorithm more robust to noise or corrupted pixels, and thus leading to improved demosaicing quality. Figure 4.7 shows the comparison of the results from dictionary learning with group sparse coding [Mairal et al., 2009b] and those from the method of Mairal et al. [2008b] (without group sparsity).

4.2.4 OTHER SIGNAL RECOVERY APPLICATIONS

In Zhang et al. [2011], the image recovery problem is considered jointly with the image classification problem, based on the sparse representation prior that, the degraded input image, if corrected recovered, should be sparsely represented by the dictionary with small residual and that sparse representation indicates the class label of the input. In contrast, other typical methods consider these two tasks separately: the degraded input image is first recovered and then the recovered image is classified. Experiments on face recognition were reported, suggesting that the recovery and classification tasks may benefit each other, when jointly modeled.

Dictionary learning has been studied for speech enhancement in Sigg et al. [2010], where two types of noise were modeled: non-structured noise **n**, e.g., Gaussian noise, which cannot be sparsely represented by a dictionary; and structured noise **i**, e.g., background music, which is sparsely representable by an inference dictionary $\mathbf{D}_i$. The clean speech signal **s** is also assumed to be sparsely representable on the speech dictionary $\mathbf{D}_s$. Mathematically, this problem can be written as:

$$\mathbf{x} = \mathbf{s} + \mathbf{i} + \mathbf{n} = \mathbf{D}_s \, {}_s + \mathbf{D}_i \, {}_i + \mathbf{n}, \tag{4.9}$$

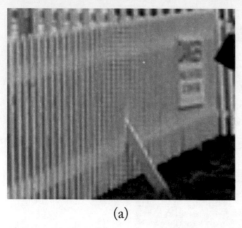
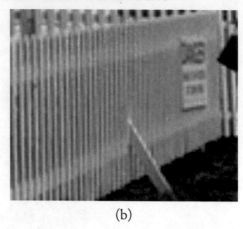

(a)   (b)

Figure 4.7: Examples of image demosaicing with algorithms proposed in Mairal et al. [2008b] (left 40.98 dB) and Mairal et al. [2009b] (right, PSNR-41.24dB) respectively. Obvious artifacts may be found in the center of the left image. These figures are from Fig. 3 of Mairal et al. [2009b].

where $\mathbf{x}$ is the input signal. The speech dictionary is learned off-line from some training data and the inference dictionary is also learned off-line from the signal segment when the speech is paused.

MR image reconstruction from partial samples was studied in Liu et al. [2013b]. Compared with other dictionary learning based image reconstruction methods, the method proposed in Liu et al. [2013b] learns a dictionary from the gradient space, i.e., the x-derivative and the y-derivative of the image, instead of from the image directly. That is, the problem is defined as

$$\mathbf{u}, \alpha, \mathbf{D} \quad : \quad \min \sum_i \|\mathbf{D}^x \alpha_i^x - R_i(\nabla_x \mathbf{u})\|_2^2 + \|\mathbf{D}^y \alpha_i^y - R_i(\nabla_y \mathbf{u})\|_2^2 + \lambda \|F(\mathbf{u}) - \mathbf{f}\|_2^2$$
$$\text{s.t.} \quad \|\alpha_i\|_0 \leq t \, \forall i, \tag{4.10}$$

where $\mathbf{f}$ is the input samples, $\mathbf{u}$ the recovered image, $F(\cdot)$ the sampling operator, $R_i(\cdot)$ the operator extracting patches at i. ∇_x and ∇_y are the derivative operators in the x and y directions, respectively, and $\mathbf{D} = \{\mathbf{D}^x, \mathbf{D}^y\}$, with the components being the dictionaries learned for patches of the x-derivative and the y-derivative, respectively. This method can be viewed as the image-adaptive version of total variation (TV) regularization, which, via modeling the gradients of the estimated image, results in better recovery quality, as shown in the examples given in Fig. 3 of Liu et al. [2013b].

4.3   IMAGE SUPER-RESOLUTION

Dictionary learning has also been successfully applied to reconstructing a high-resolution image from a low-resolution copy (or copies), namely super-resolution (SR) [Nasrollahi and Moeslund,

2014, Van Ouwerkerk, 2006]. Depending on the number of input images used, this problem can be further divided into multi-frame SR and single-frame SR. In multi-frame SR, multiple low-resolution images of the scene (with some variations such as spatial shift) are used as the input [Farsiu et al., 2004, Hardie et al., 1997]. In single-frame SR, only one low-resolution image is available as the input, which makes it a more difficult problem [Glasner et al., 2009]. Image SR can be viewed as the reversed operation of the following process:

$$L = (B * H)_{\downarrow} + n, \tag{4.11}$$

where $L \in \mathbb{R}^{w \times h}$ is the down-sampled low-resolution image, $H \in \mathbb{R}^{kw \times kh}$ is the high-resolution image, B is some blurring kernel, n is noise, k is the scale-up factor, and $(\cdot)_{\downarrow}$ is the down-sampling operator.

In SR, we want to reconstruct the high-resolution image H and possibly also the blur kernel B from the low-resolution image L. In multi-frame SR, we assume there are different blur kernels B_i which generate different low-resolution L_i from the same high-resolution H. When there are sufficiently many low-resolution images in the input, it is possible to confidently reconstruct H and also the blur kernel. However, when there are only limited number of low-resolution images available in the input, with the extreme case of only one low-resolution copy (i.e., the single-frame SR problem), this problem becomes ill-posed, and additional constraints or assumptions are often introduced to alleviate the issue. Examples include exemplar-based methods [Glasner et al., 2009] and methods imposing image priors on high-resolution images [Zhang et al., 2012a].

In Yang et al. [2008, 2010a], a single-frame SR algorithm was proposed based on a pair of low-resolution (LR) and high resolution (HR) dictionaries. The method assumes that the sparse coefficient of a patch in a high-resolution image should be unaltered after the downscaling process in Eq. 4.11, with appropriate dictionaries, i.e.,

$$D^L \alpha_i = L_i = (B * H_i)_{\downarrow} = (B * (D^H \alpha_i))_{\downarrow}, \tag{4.12}$$

where D^L and D^H are the LR-HR dictionary pair. To build such a dictionary, a coupled-dictionary learning algorithm was proposed. Given training set which contains a set of high-resolution images X and their corresponding low-resolution counterparts Y, which can be easily obtained via downsampling the high-resolution images, the algorithm learns the LR-HR dictionary pair, under which the corresponding patches from the HR images and LR images share the same sparse coefficient α

$$D^H, D^L, \alpha : \min_{D^H, D^L, \alpha} \sum_i \frac{1}{2} \|x_i - D^H \alpha_i\|_2^2 + \frac{1}{2} \|y_i - D^L \alpha_i\|_2^2 + \lambda \|\alpha_i\|_1, \tag{4.13}$$

where x_i and y_i are the patches extracted from the HR image and its corresponding LR image at the same location, respectively. To learn the dictionary pair, instead of using a large set of images, it was shown that a small set of images of similar statistical nature would serve the purpose. More details can be found in Sec. 3.3.3 or in Yang et al. [2008, 2010a].

With the learned LR-HR dictionary pair, the HR image can be reconstructed from the input LR image with the following process: extract patches from the LR image preferably with overlapping to remove boundary artifacts; for each extracted patch, apply sparse coding with the LR dictionary; reconstruct the new patch with the sparse coefficient and the HR dictionary and finally stitch the new patches to reconstruct the HR image. The algorithm is summarized in Algorithm 20.

Algorithm 20 The super resolution algorithm proposed in Yang et al. [2008, 2010a]

Input : input LR image $\mathbf{Y}$, LR-HR dictionary pair $\mathbf{D}^L$ and $\mathbf{D}^H$
Output : HR image $\mathbf{X}$

1: Extract patches $\{\mathbf{y}\}$ from LR image with overlapping
2: **for** each patch $\mathbf{y}$ **do**
3:    Compute the sparse coefficient α for $\mathbf{y}$ with the LR dictionary $\mathbf{D}^L$
4:    Reconstruct the new patch $\mathbf{x}$ as $\mathbf{D}^H \alpha$
5:    Fill the patch to the proper location of the HR image, for overlapping region use the weighted average;
6: **end for**

Similar ideas were also studied in Wang et al. [2012], where a linear projection was assumed instead of one-to-one strict correspondence between the LR dictionary and the HR dictionary; also, in Huang and Wang [2013], where the LR dictionary and the HR dictionary are first projected into a common latent subspace; and in He et al. [2013], which proposed a Bayesian method for LR-HR dictionary learning. The idea was further studied in Zeyde et al. [2012] with several improvements. Specifically, the LR image is separated into the low-frequency band and the high-frequency band, where the low-frequency band is scaled up via interpolation and only the high-frequency band is scaled up via sparse representation. This separation helps to avoid the boundary problem, as the low-frequency band is scaled up as a whole image. The idea in Yang et al. [2008] was considered in conjunction with non-local statistics in Yang et al. [2011b], where the patches are first clustered into several groups via K-means, and then dictionary learning and sparse coding are performed for each group independently under a multi-task learning framework. As a result, fewer atoms are required for each dictionary and only a lower computational cost is required. The comparison for SR results from Yang et al. [2008, 2010a, 2011b] are shown in Fig. 4.8.

In Dong et al. [2011b,c], a different idea, centralized sparse coding (CSC), was proposed for dictionary learning-based image SR. CSC estimates the blur kernel and the HR image jointly:

$$\alpha : \min_{\alpha} \|\mathbf{y} - \mathbf{HD} \circ \alpha\|_2^2 + \lambda \|\alpha\|_1 + \gamma \|\alpha - \hat{\alpha}\|_p, \tag{4.14}$$

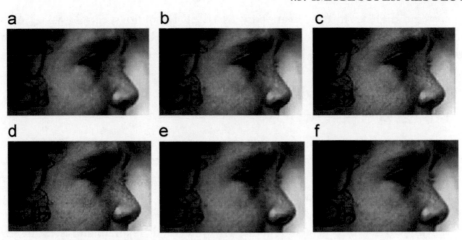

Figure 4.8: Comparisons of image super resolution results. From (a)–(f), bicubic interpolation, neighbor embedding result [Chang et al., 2004], the result of Yang et al. [2008, 2010a], the input image, and the results of Zeyde et al. [2012] with dictionary of 64 atoms and 256 atoms, respectively. The figure is from Fig. 5 of Yang et al. [2011b].

where the block diagonal matrix $\mathbf{H}$ is the projection to perform blur and downsampling, $\mathbf{y}$ is the patch from the low-resolution image, $\mathbf{D}$ is the dictionary, $\|\alpha - \hat{\alpha}\|_p$ introduces the non-local statistics such that the sparse coefficients of the patches of each group must be close to each other. It is obvious that CSC does not rely on an LR-HR dictionary pair anymore. Instead, a single HR dictionary is used, which is learned from the HR image recovered from the previous stage. For initialization, the HR image can be computed via interpolation. By iteratively refining the projection matrix $\mathbf{H}$ and the HR dictionary $\mathbf{D}$, the final HR image can be reconstructed from the dictionary and the sparse coefficients. Some sample images, showing hat CSC generate sharper edges than several other methods, can be found in Fig. 4 of Dong et al. [2011b].

In Zhang et al. [2012a], multi-scale dictionaries were explored for image SR. In the training stage, patches are extracted at multiple scales from the pyramid of training images and a multi-scale dictionary is learned from those patches. Local and non-local statistics were introduced to improve the quality of the SR image. For local prior, a pixel is assumed to be predicted as the weighted average of its neighbors, where the weight is computed via steering kernel. For non-local statistics, a non-local mean (NLM) like method was used, so that similar patches from the low-resolution image would be reconstructed to similar patches for the high-resolution image. The effects of the number of scales on the SR image quality are discussed and visualized in Zhang et al. [2012a], with comparison to other approaches.

The computational cost of SR algorithms was considered in Gao et al. [2013], where it was proposed to use efficient image interpolation for non-salient regions and dedicate the SR algorithm only for the salient regions. This method tries to achieve the best trade-off between computational cost and high SR quality. Dictionary-based SR for multi-spectral images was studied in Liu and Boufounos [2012]. In Tosic et al. [2013], image SR was studied for plenoptic imaging, based on the key idea of using dictionary learning for incoherent sampling.

4.4 SEGMENTATION

4.4.1 IMAGE SEGMENTATION

Image segmentation is the task of partitioning the pixels of an image into multiple segments, with the pixels within each segment sharing some common properties. In a basic sense, such common properties may be simple pixel features such as color or texture. In more abstract senses, such common properties may also be defined based on semantic attributes of the underlying segment. There is tremendous amount of effort on image segmentation reported in the literature. Among others, well-known algorithms include K-means, watershed [Najman and Schmitt, 1994], level set [Osher and Paragios, 2003], graph cut [Vicente et al., 2008], Grabcut [Rother et al., 2004], super pixel [Achanta et al., 2012], gPb [Arbelaez et al., 2011], and so on.

Dictionary-based approaches to image segmentation have also been proposed. In Mairal et al. [2008a, 2009c], a discriminatively-trained dictionary was used to segment an image according to texture information. For training, 12×12 patches from the training images are used to learn dictionaries, where the dictionary of each texture type has 128 atoms, with sparsity constraint $T = 4$. For testing, the sparse coefficient under a dictionary is computed for each 12×12 patch from the testing image. The label of the dictionary giving the minimal reconstruction error is used to label the patch. This initial segmentation is refined with the graph-cut algorithm based on 8 neighbors. For preprocessing, a Gaussian filter and a Laplacian filter are used for prefiltering the image. In Zhang et al. [2014], sparse learning is combined with multi-task learning to simultaneously segment multiple images that share common foreground objects. Some examples are shown in Fig. 4.9.

In Zhao et al. [2010], a special type of texture, text, was studied, where pixels belonging to text need to be segmented out. Two dictionaries are learned: the text dictionary is learned from patches with text and the background dictionary is learned from patches of natural images. In the testing stage, an SRC Wright et al. [2009c] scheme was applied to classify whether a patch containing text or not, based on the reconstruction errors under the two dictionaries. Finally, an adaptive run-length smoothing algorithm was used to refine the result. To reduce the computational cost of the testing stage, the patches are prefiltered according to its response to a wavelet-transform-based edge detector, where the patches with low responses (i.e., smooth patches) are deemed non-text. Dictionary learning has also been applied to hyperspectral image segmentation/classification in Chen et al. [2011], where contextual information was incorporated into the sparse coding problem. Specifically, a joint sparsity model was used such that the pixels within a small neighbor were encouraged to be sparsely represented by a few common dictionary atoms.

The above two methods require a training stage for texture segmentation. An unsupervised method for texture segmentation was proposed in Chi et al. [2013]. This method relies on a dictio-

Figure 4.9: Examples of image segmentation with algorithms from Zhang et al. [2014] on the MSRC dataset. In each image, the region within the green contour is the segmented foreground object.

nary learning algorithm, in which the data in the same group is assumed to share some dictionary atoms (i.e., group sparsity). This helps to improve the discriminating ability of the dictionary and the intra-block coherence of the dictionary is also minimized to reduce the redundancy of the dictionary atoms. For segmentation, an image is first divided into 32×32 overlapping regions with 16 spacing and from each region, 50 21×21 patches are randomly sampled to form a group. Dictionary learning is applied to these patches, with the dictionary divided into two sub-dictionaries, 230 atoms each. The patches from the same region are forced to only use atoms from one of the two sub-dictionaries. With the sparse coefficient computed from the dictionary learning process, the ℓ_1 norm of the coefficient of each sub-dictionary is computed and used for texture segmentation.

4.4.2   BACKGROUND SUBTRACTION

Background subtraction is a common video processing task, for which a commonly-used technique is the Gaussian mixture model [Stauffer and Grimson, 1999]. Dictionary-based representation may also be used to model the background and/or the foreground in a background subtraction problem. In Wright et al. [2009a], robust PCA (RPCA) was used for background subtraction in image sequences. The method relies on the assumption that, the background of a given sequence of images, even with changing illuminations, forms a very low-dimensional subspace. Accordingly, a matrix $\mathbf{X}$ is formed with each column $\mathbf{x}_i$ being the vectorized input images. Matrix $\mathbf{X}$ can be decomposed as $\mathbf{X} = \mathbf{A} + \mathbf{E}$, where $\mathbf{A}$ is a low-rank matrix representing the background and $\mathbf{E}$

is a sparse matrix capturing sparse noise or the foreground object. Examples of RPCA on videos are shown in Fig. 4.10.

A Bayesian interpretation of RPCA (short-handed BRPCA) was studied in Ding et al. [2011] and applied to background subtraction. This method was already described in Sec. 3.5. BRPCA generates a true low-rank matrix $\mathbf{A}$ and a true sparse matrix $\mathbf{E}$, due to the use of the Beta-Bernoulli process for $\mathbf{A}$ and $\mathbf{E}$. In contrast, in the optimization-based RPCA in Wright et al. [2009a], $\mathbf{A}$ is only approximately low-rank (most of the singular values are very close to 0) and $\mathbf{E}$ is only approximately sparse (a lot of the entries are close to 0), as the nuclear norm $\|\cdot\|_*$ and the ℓ_1 norm $\|\cdot\|_1$ are used to enforce the low-rank constraint and sparsity constraint. In addition, a Markov process was also used in Ding et al. [2011] to model the temporal relationship among the adjacent frames of a video. Experimental results (which can be found in Fig. 7 of Ding et al. [2011]) show that the method of Ding et al. [2011] outperforms RPCA Wright et al. [2009a] in background subtraction. However, both methods require the foreground object to be small. Also, they may not work well when the camera viewpoint changes quickly, as the low-rank assumption is violated.

Considering the aforementioned limitations, in Cong et al. [2011], the background $\mathbf{x}$ is assumed to be sparsely represented by a dictionary $\mathbf{D}$:

$$\alpha : \min_{\alpha} \frac{1}{2}\|\mathbf{x} - \mathbf{D}\alpha\|_1 + \lambda\|\alpha\|_1, \tag{4.15}$$

where the ℓ_1 norm is used for the reconstruction error, instead of ℓ_2 norm, because the ℓ_1 norm is more robust in modeling sparse error (e.g., a small foreground object). For a testing frame, its sparse coefficient is first computed with the current dictionary and the pixels with large reconstruction error are labeled as foreground. However, this method requires an offline training stage for the dictionary. Online dictionary learning for background subtraction was discussed in Lu et al. [2013].

4.4.3   BLIND SOURCE SEPARATION

Dictionary learning has also been applied to blind source separation, which aims to find the latent patterns in the image or other signals. In Mailhé et al. [2008], an extension to the K-SVD algorithm was proposed, which learns a shift-variant dictionary for a very long signal. The problem can be written as:

$$\mathbf{M}, \alpha \quad : \quad \min_{\mathbf{M},\alpha} \|\mathbf{x} - \sum_{k}\sum_{\tau}\alpha_{k,\tau}\mathbf{m}_k(t-\tau)\|_2^2 \tag{4.16}$$

$$: \quad \min_{\mathbf{M},\alpha} \|\mathbf{x} - \sum_{k}\sum_{\tau}\alpha_{k,\tau}T_{\tau}(\mathbf{m}_t)\|_2^2 \text{ s.t. } \|\alpha\|_0 \leq L,$$

where $\mathbf{x}$ is the input signal, $\mathbf{M}$ is the family of learned K patterns, α is the sparse coefficient, and $T_{\tau}(\cdot)$ is the shift operator that takes a pattern $\mathbf{m}$ and returns an atom that is null everywhere but for a copy of $\mathbf{m}$ that starts at time τ. A modified matching pursuit algorithm was used to compute

  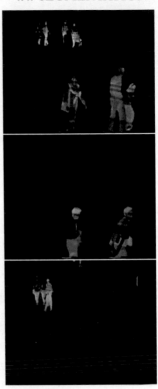

Figure 4.10: Examples of background subtraction in videos with RPCA [Wright et al., 2009a]: frames from input sequence (left), low rank matrix/background (middle) and sparse matrix/foreground (right). This figure is from Fig. 2 of Wright et al. [2009a]

the sparse coefficient and the families of patterns. The method is evaluated on extracting patterns from music tracks of 1-min long Jazz guitar pieces at 8000 Hz.

The work of Kolter et al. [2010] studied dictionary-learning-based blind source separation in the energy disaggregation problem. Energy disaggregation aims at separating the whole-home energy signal into the energy signals of the component appliances. It helps the analysis of power consumption at the device level and encourages the users to conserve energy consumption. The algorithm contains two stages: in the training stage, the dictionary is learned discriminatively from the training data of each device's power consumption over typical weeks; in the testing stage, the sparse coefficient is computed for the testing signal with the learned dictionary. Considering the non-negativity of the energy signal, the dictionary and sparse coefficient is required to be non-negative:

$$\mathbf{D}, \alpha : \min_{\mathbf{D},\alpha} \frac{1}{2}\|\mathbf{X} - \mathbf{D}\alpha\|_F^2 + \lambda\|\alpha\|_1 \text{ s.t. } \|\mathbf{d}_i\|_2 \leq 1, \mathbf{D} \geq 0, \alpha \geq 0. \tag{4.17}$$

Several extensions were also proposed to improve the performances, including total energy priors, group Lasso and shift-invariant sparse coding. For more details and experimental results, the interested readers should refer to Kolter et al. [2010].

This problem has also been studied in the image domain [Abolghasemi et al., 2012], when an image is the mixture of several image sources. The method of Abolghasemi et al. [2012] relies on the assumption that each underlying source can be sparsely represented by its corresponding dictionary. In Esser et al. [2012], blind source separation was used for unmixing hyperspectral images. The method combines the power of dictionary learning and nonnegative matrix factorization:

$$\mathbf{D}, \alpha : \min_{\mathbf{D}, \alpha} \frac{1}{2} \|\mathbf{X} - \mathbf{D}\alpha\|_F^2 + \lambda \|\alpha\|_{1,\infty} \text{ s.t. } \mathbf{D} \geq 0, \alpha \geq 0, \tag{4.18}$$

where $\ell_{1,\infty}$ norm $\|\alpha\|_{1,\infty} = \sum_i \max_j \|\alpha_{i,j}\|$ encourages the sparsity for the coefficient α. Instead of regularizing the ℓ_2 norm of the dictionary atoms (i.e., columns of $\mathbf{D}$), Esser et al. [2012] requires the columns of $\mathbf{D}$ should be selected from columns of $\mathbf{X}$.

4.5 IMAGE CLASSIFICATION

Since the initial success of sparse coding in face recognition [Wright et al., 2009c], dictionary learning for image classification has attracted a lot of attention and many newer approaches have been reported in the literature. Most approaches typically learn a dictionary $\mathbf{D}$ from the training data, and then for testing data $\mathbf{x}$, its sparse coefficient α is computed with the learned dictionary, which is used for classifying the testing sample. Based on how the sparse coefficient is used for classification, a dictionary learning-based classification approach may fall into one of the following three categories.

Using the reconstruction error. Approaches of this category (e.g., [Chen et al., 2012, Chi et al., 2013, Kong et al., 2013, Luo et al., 2013, Mairal et al., 2008a, Patel et al., 2011, 2012, Qiu et al., 2011, Zhang et al., 2013b]) assume that the dictionary atoms are attached to different class labels, e.g., let $[\mathbf{D}_1, \mathbf{D}_2, \cdots, \mathbf{D}_C]$ the set of dictionaries for Classes 1, 2,..., and C, then given the sparse coefficient α of a sample $\mathbf{x}$, the reconstruction error for atoms of Class c can be computed as $e_c = \|\mathbf{x} - \mathbf{D}_c \alpha_c\|_2$, where α_c is the subset of sparse coefficients corresponding to the subdictionary $\mathbf{D}_c$. The data $\mathbf{x}$ is then labeled to the class c^* for which the reconstruction error is minimum, i.e., $c^* = \min_c e_c$.

Using the classification score of sparse coefficient. Approaches of this category (e.g., [Jiang et al., 2011, Mairal et al., 2009c, Pham and Venkatesh, 2008, Wang et al., 2013c, Zhang and Li, 2010a]) do not require the association of dictionary atoms to different classes. Instead, a classifier $\mathbf{w}$ is learned jointly with the dictionary $\mathbf{D}$. Given the sparse coefficient α of a data $\mathbf{x}$, the classification score is computed as $\mathbf{s} = \mathbf{w}\alpha$ and the data is assigned to Class c, where s_c is maximal.

Sparse coding as nonlinear pooling operator. Instead of directly using sparse coefficients for classification, some methods use sparse coding and dictionary learning as a local feature pooling operator [Boureau et al., 2010, Coates and Ng, 2011, Feng et al., 2011a], in place of traditional vector quantization, max pooling, and sum pooling etc. The studies in Coates and Ng [2011] show that sparse coding is more effective than vector quantization in classification. It was also reported in Feng et al. [2011a] that a sparse coding based pooling scheme, where the spatial information is preserved, does better than other pooling schemes on some benchmark datasets.

4.6   SALIENCY DETECTION

An interesting and important vision task is related to how different regions or objects in the visual field capture different levels of attention of a human observer. In visual computing, this is often stated as the problem of computational visual saliency detection. One early work on this regard is the model proposed in Itti et al. [1998]. Since then, many different models have been proposed. Such models may be roughly divided into two groups: bottom-up models (stimulus-driven) that are mainly based on low-level visual features of the scene, and top-down models (goal-driven) that employ information and knowledge about a visual task. A survey of both groups of methods was reported in Borji et al. [2013]. Visual saliency analysis has been applied with success to other vision tasks including object detection [Alexe et al., 2012], image classification [Sharma et al., 2012], foreground segmentation [Li et al., 2011c], and securities [Zhao et al., 2013], etc.

In this section, we describe visual saliency detection algorithms that employ dictionary learning and sparse representation techniques. In Yan et al. [2010], image information is decomposed into two parts: redundancy, which denotes information with high regularity, and saliency, which represents novelty. More specifically, the redundancy part is assumed to be low-rank, while the saliency part prefers a small number of objects. Accordingly, this paper proposes to do saliency detection based on the robust principal component analysis (RPCA; Wright et al. [2009a]), which decomposes the images into a low-rank component and a sparse component: $X = A + E$, where X contains columns from the vectorized patches which are sampled from the given image. Instead of decomposing in the raw pixel domain, the sparse coefficients of the patches are used, where the dictionary is learned from patches of the image via the K-SVD algorithm. The saliency map is then computed via the ℓ_∞ norm of the sparse component, i.e., the saliency score for patch i is $\|e_i\|_\infty$, where e_i is the sparse component for patch i. The proposed method delivers comparable results as other the state-of-the-art visual saliency detection methods. Some examples are shown in Fig. 4.11. A similar idea has also been studied in Shen and Wu [2012].

Most existing methods consider multiple features such as color, texture, gradient, or even object/face detection, for saliency detection. In Lang et al. [2012], robust PCA is combined with multi-task learning to integrate multiple types of feature in detecting saliency. Given multiple feature matrices $\{X_i\}$, where X_i is the matrix for the i_{th} feature with one column for one patch,

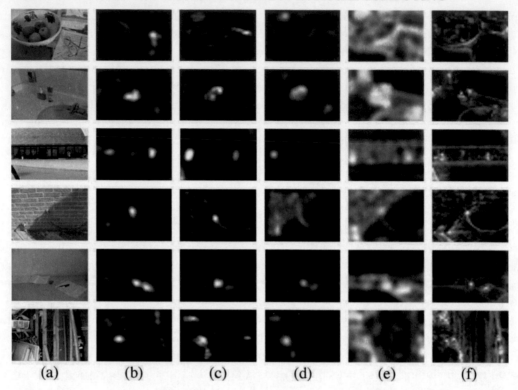

Figure 4.11: Examples of visual saliency detection by Yan et al. [2010] (Column c) with comparisons to several existing methods. (a) input image, (b) ground truth generated from human eye fixations data, and results from (c) [Yan et al., 2010], (d) [Hou and Zhang, 2009], (e) [Bruce and Tsotsos, 2005], (f) [Itti et al., 1998]. This figure is from Fig. 4 of Yan et al. [2010].

the following decomposition is computed:

$$\{\mathbf{Z}_i\}, \{\mathbf{E}_i\} : \min_{\{\mathbf{Z}_i\}, \{\mathbf{E}_i\}} \lambda \|\mathbf{E}\|_{2,1} + \sum_i \|\mathbf{Z}\|_* \text{ s.t. } \mathbf{X}_i = \mathbf{E}_i + \mathbf{Z} \forall i, \tag{4.19}$$

where $\mathbf{E} = [\text{vec}(\mathbf{E}_1), \text{vec}(\mathbf{E}_2), \cdots, \text{vec}(\mathbf{E}_K)]$. The saliency score for a patch can be computed as $\mathbf{s}_i = \sum_k \|\mathbf{E}_j[:, i]\|_2$, where $\|\mathbf{E}_k[:, i]\|_2$ is the ℓ_2 norm of the i_{th} column of $\mathbf{E}_k$. Experiments on the Bruce dataset and the MSRA dataset show promising results, with several types of features used: RGB, RGB histograms, responses of three-scale steerable filter at four orientations and local contrast.

Although RPCA-based approaches were reported to achieve promising results in visual saliency detection, the computational cost of RPCA can be very high, when the input is at high resolution. To overcome this limitation, an approach was proposed in Li and Haupt [2014] for RPCA on very large matrices, via randomized adaptive compressive sampling. In this method, a smaller measurement $\mathbf{Y}$ is collected from the input $\mathbf{X}$:

$$\mathbf{Y} = \Phi\mathbf{XS}, \tag{4.20}$$

where Φ is the measurement matrix randomly sampled from the Gaussian distribution and $\mathbf{S}$ is the column sampling matrix randomly drawn from the Bernoulli distribution. RPCA is performed on these collected measurements:

$$\mathbf{A}, \mathbf{E} : \min \|\mathbf{A}\|_* + \lambda\|\mathbf{E}\|_{1,2} \text{ s.t. } \mathbf{Y} = \mathbf{A} + \mathbf{E}. \tag{4.21}$$

At the second stage, a sparse matrix $\tilde{\mathbf{E}}$ is recovered by solving the problem $\tilde{\mathbf{E}} : \min \|\tilde{\mathbf{E}}\|_1$, s.t., $\tilde{\mathbf{E}}\Psi^T = \phi\mathcal{P}_\mathbf{A}\Phi\mathbf{X}\Psi^T$, where Ψ is the measurement matrix and ϕ is the measurement vector, randomly drawn from the Gaussian distribution. It was proven in Li and Haupt [2014] that, under certain condition (detailed in Sec. II of Li and Haupt [2014]), the sparse matrix can be recovered with high probability. Experimental results (which can be found in Fig. 4 of Yang and Yang [2012]) show that with as less as 1.5% measurement from an 300×400 image, the above method is able to detect the saliency map with accuracy comparable to other existing methods.

In Yang and Yang [2012], visual saliency detection is solved by combining conditional random field (CRF) and discriminative dictionary learning. A local image patch is modeled as the node of the field, with pair-wise edges indicating the spatial adjacency. The task is to assign binary label y to the field, where $y = 1$ indicates the patch is saliency or otherwise being just the non-salient background. The energy function of the random field can be written as:

$$E(\alpha, \mathbf{y}, \mathbf{w}, \mathbf{D}) = \sum_i \psi(\alpha_i, y_i, \mathbf{w}_1) + \sum_{i,j \subset c} \phi(y_i, y_j, \mathbf{w}_2) \tag{4.22}$$

$$= \sum_i -y_i\mathbf{w}_1^T\alpha_i + \sum_{i,j \in \epsilon} \mathbf{w}_2\delta(y_i - y_j), \tag{4.23}$$

where $(\mathbf{w}_1, \mathbf{w}_2)$ is the parameter of the CRF. Due to the semantic and geometric ambiguity of a patch under raw intensity representation, the sparse coefficient of the patch is used instead, i.e., $\alpha_i : \min_{\alpha_i} \frac{1}{2}\|\mathbf{x}_i - \mathbf{D}\alpha\|_2^2 + \lambda\|\alpha_i\|_1$. The dictionary and the CRF is learned jointly.

Human fixation data could be leveraged for supervised saliency detection. In Jiang et al. [2013], label consistent K-SVD (LC-KSVD; Jiang et al. [2011]) was used to learn a discriminative dictionary and a classifier for saliency detection:

$$\mathbf{D}, \alpha, \mathbf{A}, \mathbf{w} : \min \|\mathbf{X} - \mathbf{D}\alpha\|_F^2 + \lambda\|\mathbf{U} - \mathbf{A}\alpha\|_F^2 + \eta\|\mathbf{y} - \mathbf{w}^T\alpha\|_2^2 \text{ s.t. } \|\alpha_i\|_0 \leq T, \tag{4.24}$$

where $\mathbf{D}$ is the dictionary, α the sparse coefficient, $\mathbf{y}$ the ground truth saliency score for the training patch and $\mathbf{w}$ the classifier for assigning a saliency score with the sparse coefficient. $\|\mathbf{U} - \mathbf{A}\alpha\|_F^2$ enforces label consistency. Examples of the learned dictionaries for both salient patches and non-salient patches at different scales are presented in Fig. 2 of Jiang et al. [2013].

Saliency detection has also been studied in the video domain [Zhao et al., 2011], where the aim is to detect unusual "events" from a video. A given video sequence is divided into many spatiotemporal cuboids $\mathbf{x}_i^j$ and unusual events $\mathbf{X}_i$ correspond to some cuboids. To locate the cuboids belonging to the unusual events, the cuboids are sparsely coded and their reconstruction error are then used as a measurement of being unusual:

$$J(\mathbf{X}_i, \alpha_i, \mathbf{D}) = \frac{1}{n_i} \sum_j \frac{1}{2} \|\mathbf{x}_i^j - \mathbf{D}\alpha_i^j\|_2^2 + \lambda \|\alpha_i^j\|_1 + \frac{\eta}{n_i} \sum_k w_{j,k} \|\alpha_i^j - \alpha_i^k\|_2^2, \qquad (4.25)$$

where the second term enforces smoothness that the adjacent cuboids having similar sparse coefficient with $w_{j,k}$ related to the distances of cuboids j and k in the spatiotemporal domain. An event $\mathbf{X}_i$ is marked to be unusual, if $J(\mathbf{X}_i, \alpha_i, \mathbf{D}) \geq \epsilon$, where ϵ is some predefined constant.

Besides images and videos, saliency detection has been applied in other contexts. In Kasiviswanathan et al. [2011], novel topics in the text are viewed as salient, which cannot be sparse-represented by a dictionary learned from the text. As a result, the novelty of the document is measured as its reconstruction error over the dictionary. To scale up the algorithm for social media data, e.g., Twitter stream, an online algorithm was proposed.

4.7   VISUAL TRACKING

Visual tracking, a common task in video-related visual computing, has seen great applications of dictionary learning and sparse representation. An early attempt can be traced to Mei and Ling [2009], where visual tracking is formulated as a sparse approximation problem in a particle filter framework. Specifically, the correct target in a new frame $\mathbf{x}$ is expected to be sparsely represented by a dictionary of the target template $\mathbf{D}$, i.e., $\mathbf{x} = \mathbf{D}\alpha + \epsilon$ with $\|\alpha\|_0 \leq T$ and the noise $\epsilon \sim \mathrm{N}(0, \sigma)$. To deal with occlusion, corruption and other challenges, a sparse residual term $\mathbf{e}$ is also introduced, leading to $\mathbf{x} = \mathbf{D}\alpha + \mathbf{e} + \epsilon$, where $\|\mathbf{e}\|_1$ is small. Accordingly, the sparse coding problem can be written as:

$$\alpha, \mathbf{e} \quad : \quad \min \|\mathbf{x} - \mathbf{D}\alpha - \mathbf{e}\|_2^2 + \lambda \|\alpha\|_1 + \lambda \|\mathbf{e}\|_1$$

$$: \quad \min \left\|\mathbf{x} - \begin{bmatrix} \mathbf{D} & \mathbf{I} \end{bmatrix} \begin{bmatrix} \alpha \\ \mathbf{e} \end{bmatrix}\right\|_2^2 + \lambda \left\|\begin{bmatrix} \alpha \\ \mathbf{e} \end{bmatrix}\right\|_1 \qquad (4.26)$$

$$\text{s.t.} \quad \alpha \geq 0.$$

The above problem can be solved by linear programming, where the nonnegativity constraint on the sparse coefficient is used to avoid the trivial tracking result. Then the tracked target is identified by checking the reconstruction error $\|\mathbf{x} - \mathbf{D}\alpha\|_2$: the correct target should have the minimum reconstruction error.

To handle the variations of target appearance in the video, the dictionary $\mathbf{D}$ for the template is dynamically updated through the tracking process in Mei and Ling [2009]: the dictionary atoms are updated according to their usage in sparse coding. More specifically, if the current tracking result is found to be significantly different from the atoms in the dictionary ($\max_k \mathrm{sim}(\mathbf{x}, \mathbf{d}_k) \leq \tau$), the least important atom $\mathbf{d}_j$, where $j : \min_k \|\mathbf{d}_k\|_2 e^{\alpha_k}$, will be replaced by the current tracking result $\mathbf{x}$. For initialization, windows sampled around the manually selected target at the first frame

---

Algorithm 21 The ℓ_1 tracking algorithm Mei and Ling [2009]

Input : video $\mathbf{v}$, target at the first frame $\mathbf{x}_1$ and parameter τ, λ
Output : tracking result $\{\mathbf{x}_t\}$ and optional dictionary $\mathbf{D}$

1: Initialize $\mathbf{D}$ from the first frame and intialize the weight of dictionary atom $\mathbf{w}$ to $\frac{1}{K}$, where K is the number of atoms in the dictionary
2: **while** having new frame **do**
3:     Draw the new frame from the video and sample candidate window from the new frame according to the particle filter
4:     **for** each candidate window **do**
5:         Apply sparse coding with Eq. 4.26 and the compute the reconstruction error
6:     **end for**
7:     Add the window with minimum reconstruction error to the tracking result $\mathbf{x}_t$
8:     Update the weight of the dictionary atom as $\mathbf{w} = \mathbf{w} \circ e^{\alpha}$
9:     **if** $\max_k \mathrm{sim}(\mathbf{x}_t, \mathbf{d}_k) \leq \tau$ **then**
10:         Replace the atom $\mathbf{d}_j$ by $\mathbf{x}_t$, with $j : \min_k \|\mathbf{d}_k\|_2 e^{\alpha_k}$
11:         Update the weight for the new atom as median($\mathbf{w}$)
12:     **end if**
13:     Normalize the weight $\mathbf{w}$ such that $1^T\mathbf{w} = 1$ and adjust $\mathbf{w}$ such the its maximum is no larger than 0.3
14:     Normalize the dictionary atom such that $\|\mathbf{d}_k\|_2 = \mathbf{w}_k$
15:     Update the particle filter
16: **end while**

---

are used as the atoms of the dictionary, after being normalized to zero mean and unit ℓ_2 norm. Particle filtering is used to sample the candidate window of the target in the new frame, where an affine model is assumed for target motion, with six parameters in the state vector.

The potential advantages of the above ℓ_1 tracker include robustness to occlusion and appearance change. The tracking algorithm is summarized in Algorithm 21. Sample tracking results from Mei and Ling [2009] with comparisons to three other methods are shown in Fig. 4.12.

This tracking approach has also been applied to vehicle tracking in infrared videos Ling et al. [2010]. In Tzimiropoulos et al. [2011], it was found that, compared with raw pixel intensity, gradient orientation is more robust for visual tracking, especially under occlusion. Region covariance descriptor, $C(x, y) = \frac{1}{N-1} \sum_{(i,j) \in N(x,y)} (f_{i,j} - \mu)(f_{i,j} - \mu)^T$, which is the covariance matrix of a local patch, is also considered for visual tracking in Wu et al. [2011]. By applying log-transform, the covariance matrix can be measured in the Euclidean space instead of the Riemannian manifold and accordingly, the ℓ_1 tracker can be applied. The region covariance descriptor

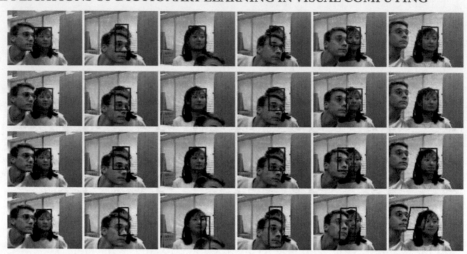

Figure 4.12: Examples of visual tracking, where the tracking results are indicated by red boxes. From top to bottom, results of Mei and Ling [2009], Comaniciu et al. [2003], Porikli et al. [2006], and Zhou et al. [2004]. This figure is from Fig. 7 of Mei and Ling [2009]

is also capable of supporting fusion of different types of features such as color and gradient. Huber loss $\ell_\lambda(r) = \begin{cases} \frac{1}{2}r^2 & \text{if } |r| < \lambda \\ \lambda|r| - \frac{1}{2}\lambda^2 & \text{otherwise} \end{cases}$ was introduced for computing the reconstruction error in sparse coding, which is presumably more robust to occlusion and outliers than the ℓ_2 norm. Huber loss acts like the ℓ_2 norm when the input is small, but will penalize linearly as the input gets larger. In addition, both the dictionary atoms and the sparse coefficients are required to be non-negative. Accordingly, the objective function can be written as:

$$\alpha, \mathbf{D} : \min \ell_\lambda(\mathbf{x} - \mathbf{D}\alpha) + \gamma \|\alpha\|_1 \text{ s.t. } \alpha \geq 0, \mathbf{D} \geq 0. \tag{4.27}$$

It was reported that experimental results demonstrated that, by using the Huber loss and the non-negative constraint on both the dictionary atoms and the coefficients, the above tracking algorithm outperforms several other existing trackers.

To handle tracking drift, which occurs in many template-based tracking algorithms, Xing et al. [2013] proposed to learn three dictionaries, each having a different lifespan: the dictionary with a short lifespan is learned only from the previous frame to capture the most recent variation and the dictionary with a long lifespan is learned from all past frames. To further improve the performance, a negative dictionary is also learned from the off-target region of previous frames.

Online dictionary learning was used in Bai and Li [2012], which progressively updates the dictionary with the following formulation:

$$\mathbf{D}, \alpha : \min \frac{1}{t} \sum_i \|\mathbf{x}_i - \mathbf{D}\alpha_i\|_2^2 + \lambda \|\alpha_i\|_1 \text{ s.t. } \|\mathbf{d}_k\|_2^2 + \gamma \|\mathbf{d}_k\|_1 \leq 1 \forall k. \tag{4.28}$$

The elastic constraint $\|\mathbf{d}_k\|_2^2 + \gamma\|\mathbf{d}_k\|_1 \leq 1$ is used due to its capability of preserving the entries in the dictionary which can better capture the local appearance of a target. In addition, a robust similarity metric is introduced to identify and reject outliers and corrupted pixels, which is defined as $\sqrt{\frac{\sum_i w_i r_i}{\sum_i}}$, where r_i is the reconstruction error for the i_{th} pixel, σ is median absolute deviation of $\mathbf{r}$, and $\mathbf{w}$ is computed as:

$$w_i = \begin{cases} 1 & \text{if } |\frac{r_i}{\sigma}| \leq \tau \\ 0 & \text{otherwise.} \end{cases} \tag{4.29}$$

Experimental results show that the method outperformed several existing trackers for some very challenging video sequences (some such examples can be found in Fig. 4 of Bai and Li [2012]).

In visual tracking, often several candidate windows need to be evaluated, which can be sampled according to the distribution computed via a particle filter framework. Considering possible intrinsic relationship among these candidate windows, Zhang et al. [2012c] proposes to introduce multi-task learning into sparse-coding-based visual tracking, where group sparse coding is used to find the sparse representations for all candidate windows simultaneously. This encourages the candidate windows to share similar dictionary atoms. More specifically, the $\ell_{p,q}$ mixture norm is used for this purpose:

$$\alpha : \min \|\mathbf{X} - \mathbf{D}\alpha\|_F^2 + \lambda\|\alpha\|_{p,q}. \tag{4.30}$$

Typically, $q = 1$ is used with $p = 2$, or $q = 1$ and $p = \infty$. Experimental results of Zhang et al. [2012c] show that by imposing group sparsity via the $\ell_{p,q}$ mixture norm, the tracking performance is improved over the existing ℓ_1 tracker and other trackers. Similar idea was also studied in Zhang et al. [2013c].

The ℓ_1 based technique can still be too computationally expensive for real-time tracking, especially if a lot of candidate windows need to be evaluated. Several methods have been proposed to reduce the computational cost of sparse coding in tracking. In Liu et al. [2010], the computational cost is reduced by selecting a subset of features for sparse coding. It was found that, only a small subset of features is useful in visual tracking. To this end, a feature selector $\mathbf{w}$ is used, where $w_i \neq 0$ if i_{th} feature is used, which can be learned from the training data. The sparse coding problem is then formulated as:

$$\alpha, \mathbf{e} : \min \|\text{diag}(\mathbf{w})\mathbf{x} - \text{diag}(\mathbf{w})\mathbf{D}\alpha - \text{diag}(\mathbf{w})\mathbf{e}\|_2^2 + \lambda\|\alpha\|_1 + \lambda\|\mathbf{e}\|_1. \tag{4.31}$$

Experimental results show that, even with a subset of features, the method proposed in Liu et al. [2010] is still able to achieve comparable or even better result than other ℓ_1 based tracking algorithms. Similar idea was studied in Wu et al. [2011], where spatial pyramid was also applied to incorporate the spatial information into the template.

Instead of learning a feature selector to select a small subset of features for sparse coding, a random measurement matrix can be used instead, as done in Li et al. [2011a], Mei and Ling [2009], and Zhang et al. [2012b]. These are related to the concept of compressive sensing.

The work in Liu et al. [2011b] focuses on reducing the computational cost of sparse approximation in tracking by gradually optimizing the sparsity and reconstruction error. In Bao et al. [2012], accelerated proximate gradient is used for solving the sparse coding problem, which delivers guaranteed quadratic convergence speed. In Mei et al. [2011], the computational cost is reduced by screening out the candidates samples before applying sparse coding, which is achieved by reconstruction without the sparsity constraint, where the candidates with large reconstruction error are rejected from further sparse coding.

While most of the aforementioned sparse representation based tracking algorithms rely on the assumption that the correct target in the new frame should be sparsely represented by the learned dictionary, Zhang et al. [2010] proposed to find the correct target in a "reversed" way. In this method, a dictionary is built with atoms being the candidate windows of the new frame and the target template is sparsely coded over the dictionary. The idea assumes that, in the sparse coefficient, the atom corresponding to the correct target in the new frame should be non-zero. Accordingly, the tracking result in the new frame can be identified by tracking the index of the non-zero element in the sparse coefficient. Experimental results show that this method outperforms the mean-shift-based tracking algorithm on several challenging video sequences.

CHAPTER 5

An Instructive Case Study with Face Recognition

Sparse representation based on data-adaptive dictionary learning has evolved into a powerful, computational scheme, with different algorithms developed for various types of applications. The examples and discussion in the previous chapters only serve to illustrate a fraction of the diversity and general applicability of sparse representation and its various algorithms. In this chapter, we use a well-studied application, face recognition, as a case study to illustrate the general design strategies in applying the dictionary-learning-based sparse techniques. This is largely based on the approaches reported in Wright et al. [2009c] and Zhang and Li [2010a]. Such an instructive case study will serve to demonstrate the following typical tasks one would often need to address in building a dictionary-learning-based solution to a given problem: coming up with a proper dictionary-based formulation, finding a solution (or an approximate solution) for the learning task under the formulation, understanding the behavior of the the solution (e.g., convergence analysis of an optimization algorithm), developing inference schemes such as classification (if needed) under the learned dictionary, and extending a basic model by incorporating additional constraints specific to a given problem, etc.

Being able to recognize faces is an importance component of human intelligence. In this regard, computer vision has yet to catch up with human performance especially under unconstrained acquisition conditions. Depending on application-specific assumptions, there may be many slightly different ways of defining the task of face recognition by a computer. A basic version can be simply given as: Given a set of subjects with one or more face images for each of them, to design an algorithm to recognize a new face image as being from one of the subjects in the given set (and possibly with the capability to reject the new image, i.e., declaring that it does not come from any subject in the given set).

More formally, let $X_i = [x_{i,1}, x_{i,2}, \ldots, x_{i,n_i}] \in R^{m*n_i}$ be the matrix representing the n_i face images from the i-th person, where each $x_{i,j}$ is an m-dimensional vector formed by vectorizing the j-th image from the i-th person. The entire given set of images can then be written as the following big matrix:

$$X = [X_1, X_2, \ldots, X_k] \tag{5.1}$$

with k being the number of subjects in the set. Then the above basic face recognition task becomes finding a label (from 1 to k) or rendering a rejection for any new image y, given the above data matrix X.

Over the years, many approaches have been proposed to solve the above task. Among others, the early and probably most well-known approaches are the Eigenface technique [Turk and Pentland, 1991] and the Fisherface approach [Belhumeur et al., 1997]. Recent years continue to see new approaches being proposed. In the rest of this chapter, we illustrate how this task may be addressed by approaches based on dictionary learning and sparse representation.

5.1   A BASIC DICTIONARY-BASED FORMULATION

Owing to the special structure of the human face, a vectorized face image $\mathbf{y}$ cannot lie in any random location in the m-dimensional space. The Eigenface technique effectively finds a lower-dimensional subspace through eigen decomposition of the training face images. In a dictionary-based scheme, the basic idea is to view the set of training face images $\{\mathbf{x}_{i,j}\}$ as some defacto "bases," hoping they could serve to somehow "span" the sub-space of all face images under consideration. In practice, if we assume that images from the same subject should be more closely related to each other than to images from a different subject, we effectively assume some sparsity of the representation of a face image under the above set of defacto bases. Formally, the above idea may be expressed by representing a face image $\mathbf{y}$ as

$$\mathbf{y} = \mathbf{X}\alpha \qquad (5.2)$$

with $\|\alpha\|_p$ (with $p = 0$ or $p = 1$) being small to reflect the sparsity constraint.

Note that the above assumptions also allude that the non-zero entries of α should concentrate on those training images in $\mathbf{X}$ that belong to the same subject of $\mathbf{y}$. In practice, considering that the representation of any face image under the training data matrix is only an approximation, we may implement the above basic idea based on solving the following sparse coding problem:

$$\alpha = \underset{\alpha}{\mathrm{argmin}} \|\mathbf{y} - \mathbf{X}\alpha\|_2 \text{ s.t. } \|\alpha\|_0 \leq T. \qquad (5.3)$$

Here we use $p = 0$, but $p = 1$ would also serve the purpose. To solve the problem in Eq. 5.3, we can use Orthonormal Matching Pursuit [Pati et al., 1993, Tropp and Gilbert, 2007] ($p = 0$) or GPSR [Figueiredo et al., 2007] ($p = 1$). This is in fact the basic idea behind the SRC approach [Wright et al., 2009c], although to make the idea practical, more tricks should be introduced, such as relying on the reconstruction error for classification (rather than counting where the non-zero coefficients come from).

5.2   AN IMPROVED FORMULATION

In the basic formulation of the previous section, the raw data matrix $\mathbf{X}$ may be only sub-optimal for supporting the classification task as the raw images are simply put together without attempting to use them discriminatively. Besides, the matrix may grow too big (for a large face image set) to accommodate efficient processing. To improve upon the basic formulation, one may first introduce dictionary learning to reduce the size of $\mathbf{X}$ while keeping its representational power. This problem can be written as:

$$\mathbf{D}, \alpha \min_{\mathbf{D},\alpha} \|\mathbf{X} - \mathbf{D}\alpha\|_F^2 \text{ s.t. } \|\alpha_i\|_0 \leq T \forall i \tag{5.4}$$

which is essentially the K-SVD algorithm that learns a compact dictionary $\mathbf{D}$ to sparsely represent all the training samples $\mathbf{x}_{i,1}, \cdots, \mathbf{x}_{k,n_k}$.

With the learned dictionary $\mathbf{D}$, we could find the sparse coefficient of a sample via Eq. 5.3. However, we are no longer able to label an image by the concentration of the non-zero entries in α, as during dictionary learning in Eq. 5.4, there is no label information attached to the atoms of the dictionary. Thus, we will need a classifier to infer the label from the sparse coefficient of an input. In the case of linear classification with a squared loss function (or more complex models, e.g., logistic loss [Mairal et al., 2009c], which is, however, not covered here), we have:

$$\mathbf{W} : \min_{\mathbf{w}} \|\mathbf{H} - \mathbf{W}\alpha\|_F^2 \text{ s.t. } \|\mathbf{W}\|_F^2 \leq \tau, \tag{5.5}$$

where $\mathbf{W}$ is the parameter for the linear classifier, $\mathbf{H}$ is the label vector for the training samples, in which $h_i^j = 1$ indicates the i-th sample takes Label j otherwise $h_i^j = 0$, and $\|\mathbf{W}\|_F^2 \leq \tau$ ensures that there is no over-fitting in $\mathbf{W}$. In practice, Eq. 5.4 and Eq. 5.5 are alternated for several iterations to seek better performance [Pham and Venkatesh, 2008], which is essentially equivalent to solving the following problem:

$$\mathbf{D}, \mathbf{W}, \alpha \; : \; \min_{\mathbf{D},\mathbf{W},\alpha} \|\mathbf{X} - \mathbf{D}\alpha\|_F^2 + \gamma\|\mathbf{H} - \mathbf{W}\alpha\|_F^2 + \beta\|\mathbf{W}\|_F^2$$
$$\text{s.t.} \quad \|\alpha\|_0 \leq T, \tag{5.6}$$

where γ and β are scalars controlling the relative contribution of the terms to the overall objective function.

5.3 SOLVING THE LEARNING PROBLEM

The iterative algorithm mentioned above can only find an approximate solution to the problem of Eq. 5.6, since in each step of the method, it only finds solution for a sub-problem of Eq. 5.6, i.e., Eq. 5.4 and Eq. 5.5. While practically speaking, the final solution may converge to the real solution, the method has high risk of getting stuck at local minima of the subproblems. Also, the convergence speed can be very slow. To get around these issues, and to leverage the proven performance of the K-SVD algorithm, we may employ the following Discriminative K-SVD (D-KSVD) algorithm, which utilizes K-SVD to find the globally optimal solution for all the parameters simultaneously. In D-KSVD, the task is formulated as solving the following problem:

$$\mathbf{D}, \mathbf{W}, \alpha : \min_{\mathbf{D},\mathbf{W},\alpha} \left\| \begin{pmatrix} \mathbf{X} \\ \sqrt{\gamma}\mathbf{H} \end{pmatrix} - \begin{pmatrix} \mathbf{D} \\ \sqrt{\gamma}\mathbf{W} \end{pmatrix} \alpha \right\|_F^2 + \beta\|\mathbf{W}\|_F^2 \text{ s.t. } \|\alpha\|_0 \leq T. \tag{5.7}$$

We adopt the protocol in the original K-SVD algorithm: the matrix $\begin{pmatrix} \mathbf{D} \\ \sqrt{\gamma}\mathbf{W} \end{pmatrix}$ is always normalized column-wise. Therefore, we can further drop the regularization penalty term $\|\mathbf{W}\|_F^2$, and thus the final formulation of the problem can be written as:

$$\mathbf{D}, \mathbf{W}, \alpha : \min_{\mathbf{D}, \mathbf{W}, \alpha} \left\| \begin{pmatrix} \mathbf{X} \\ \sqrt{\gamma}\mathbf{H} \end{pmatrix} - \begin{pmatrix} \mathbf{D} \\ \sqrt{\gamma}\mathbf{W} \end{pmatrix} * \alpha \right\|_F^2 \text{ s.t. } \|\alpha\|_0 \leq T. \tag{5.8}$$

Now, the problem of Eq. 5.8 can be efficiently solved by updating the dictionary atom by atom with the following method: For each atom $\mathbf{d}_k$ and the corresponding coefficient α_k, we solve the following problem:

$$\mathbf{d}_k, \alpha_k : \min_{\mathbf{d}_k, \alpha^k} \|\mathbf{E}_k - \mathbf{d}_k \alpha^k\|_F, \tag{5.9}$$

where $\mathbf{E}_k = \mathbf{X} - \sum_{i \neq k} \mathbf{d}_i \alpha^i$. This is essentially the same problem that K-SVD has solved and thus the solution to Eq. 5.9 is given by

$$\begin{aligned} \mathbf{U}\Sigma\mathbf{V}^T &= \text{SVD}(\mathbf{E}_k) \\ \tilde{\mathbf{d}}_k &= \mathbf{U}_1 \\ \tilde{\alpha}_k &= \Sigma(1,1)\mathbf{V}_1, \end{aligned} \tag{5.10}$$

where $\mathbf{U}_1$ denotes the first column of U. The convergence of the algorithm is guaranteed by the convergence behavior of the K-SVD algorithm.

Upon the completion of training with the labeled data in the above D-KSVD algorithm, we obtain an learned dictionary $\mathbf{D}$ and a classifier $\mathbf{W}$. However, the dictionary $\mathbf{D}$ does not readily support a sparse-coding based representation of a new test image, since $\mathbf{D}$ and $\mathbf{W}$ are normalized jointly in the previous learning algorithm, i.e,

$$\left\| \begin{pmatrix} \mathbf{d}_i \\ \sqrt{\gamma}w_i \end{pmatrix} \right\|_2 = 1. \tag{5.11}$$

Note that we cannot simply re-normalize $\mathbf{D}$ column-wise by itself, since in the training stage $\mathbf{W}$ is obtained with the original, un-normalized $\mathbf{D}$. Instead, the normalized dictionary $\mathbf{D}'$ and the corresponding classifier $\mathbf{W}'$ can be computed as

$$\begin{aligned} \mathbf{D}' &= \{\mathbf{d}'_1, \mathbf{d}'_2, \ldots, \mathbf{d}'_k\} = \{\frac{\mathbf{d}_1}{\|\mathbf{d}_1\|_2}, \frac{\mathbf{d}_2}{\|\mathbf{d}_2\|_2}, \ldots, \frac{\mathbf{d}_k}{\|\mathbf{d}_k\|_2}\} \\ \mathbf{W}' &= \{\mathbf{w}'_1, \mathbf{w}'_2, \ldots, \mathbf{w}'_k\} = \{\frac{\mathbf{w}_1}{\|\mathbf{d}_1\|_2}, \frac{\mathbf{w}_2}{\|\mathbf{d}_2\|_2}, \ldots, \frac{\mathbf{w}_k}{\|\mathbf{d}_k\|_2}\}. \end{aligned} \tag{5.12}$$

5.4   FACE RECOGNITION WITH THE LEARNED DICTIONARY

With the normalized $\mathbf{D}'$, we find the sparse coefficient for the given sample. We are now ready to assign a label to a test sample according to its sparse coefficient, for which we can use:

$$l : \max_l (\mathbf{W}\alpha)_l, \qquad (5.13)$$

where $(\cdot)_i$ refers to the i-th element.

Note that the coefficient α' can be viewed as the weight of each atom in reconstructing the test image. Thus, we can view each column $\mathbf{w}'_k$ of $\mathbf{W}'$ as a factor for measuring the similarity of atom $\mathbf{d}'_k$ to each class. Therefore, $l = \mathbf{W}' * \alpha'$ is the weighted similarity of the test image $\mathbf{x}$ to each class. In this sense, the label of test image y is decided by the index i where l_i is the largest among all elements of the l computed in Eq. 5.13. Obviously, in the ideal case, l will be of the form $l = \{0, 0, \ldots, 1, \ldots, 0, 0\}$ (i.e, with only one non-zero entry, which equals to 1).

The above dictionary-learning-based approach has been found to be very effective [Zhang and Li, 2010a]. For completeness of discussion, we now recap the major experiments and results that were reported therein, so that an interested reader can immediately gain a general sense of the performance of the approach. This also serves to illustrate a few details that one would need to consider in implementing such types of techniques.

The experiments were based on the YaleB dataset, which contains about 2414 frontal face images of 38 individuals. Similar to what was done in Wright et al. [2009c], cropped and normalized face images of 192×168 pixels were used, which were taken under varying illumination conditions. The dataset was randomly split into two halves. One half, which contains 32 images for each person, was used for training the dictionary. The other half was used for testing. Further, face images $\in \mathbb{R}^{192*168}$ were mapped into vectors $\in \mathbb{R}^{504}$ with a randomly generated matrix (the same idea of Randomface Wright et al. [2009c]). The learned dictionary contains 304 atoms, which corresponds to, on average, roughly 8 atoms for each person. The caveat is, unlike in the SRC algorithm, in D-KSVD there is no explicit correspondence between the atoms and the labels of the subjects, since all the information is encapsulated into the discriminative dictionary and the corresponding classifier. The sparsity prior assumed in the learning was set to $T = 16$.

With the above protocol, three methods—SRC, a baseline algorithm (the method of Pham and Venkatesh [2008]), and the D-KSVD method—were compared. The best result reported by SRC is 98.26% when there are 32 images per person in the dictionary. However, the performance of SRC suffers when the dictionary is smaller (8 atoms per people), which is denoted by $SRC\dagger$ in the subsequent tables. In Pham and Venkatesh [2008], the authors only used a few images (at most 4) per person for training and the recognition result was very poor (about 66.4%). For a fair comparison, more training images were used for the baseline algorithm. This key learning parameter was kept the same for all three methods. Table 5.1 summarizes the major results.

In addition to classification accuracy, the speed performance in testing a new image was also compared. This was done by recording the total time for classifying all the test images, and then

Table 5.1: The performance of the algorithms (recognition rate in %) for the extended YaleB database. The images for training the dictionary were randomly selected

D-KSVD	SRC	SRC†	Baseline
95.56	99.31	80.76	93.17

Table 5.2: The time for classifying one test image using the SRC method and the D-KSVD method on the extended YaleB database. We record the time for all the test images and then divide it by the number of images. The value is the average over 4 rounds. The unit is millisecond

Method	D-KSVD	SRC	SRC†
Time	84	120	83

divided it by the number of the test images, hence obtaining the average processing time for each testing image. Four rounds of such experiments were averaged to generate the final number, as shown in Table 5.2. From the results in Table 5.2, we can see that, with a smaller dictionary (304 atoms in the dictionary for D-KSVD and SRC†, and 1216 atoms in the dictionary for SRC), we can save about 1/3 of the time in testing. With a database involving more people, we can expect that a smaller dictionary can save even more, making the method practical for applications involving large-scale data.

To conclude the discussion on the case study, we call the reader's attention again to the classification scheme given by Eq. 5.13, which employs both W and α, with the latter been solved based on the learned dictionary $\mathbf{D}'$. Despite the notational simplicity of this classification scheme, the discussion around it and its good performance clearly demonstrates that the learning algorithm has led us to a compact dictionary that is information-laden for our analysis task. In particular, each atom of the learned dictionary encapsulates a lot more information than those in the basic dictionary formed by putting together the raw images as in Eq. 5.1. Practically speaking, this is the goal of many of the algorithms presented in earlier chapters. Clearly, with the dictionary atoms encoding presumably much richer information than a raw data sample, it is not surprising that classification now needs to be done with a scheme much more sophisticated than simple association.

Bibliography

Abolghasemi, V., Ferdowsi, S., Makkiabadi, B., and Sanei, S. (2012). Adaptive fusion of dictionary learning and multichannel bss. In *Acoustics, Speech and Signal Processing (ICASSP), 2012 IEEE International Conference on*, pages 2421–2424. IEEE. DOI: 10.1109/ICASSP.2012.6288404. 44, 94

Achanta, R., Shaji, A., Smith, K., Lucchi, A., Fua, P., and Susstrunk, S. (2012). Slic superpixels compared to state-of-the-art superpixel methods. *Pattern Analysis and Machine Intelligence, IEEE Transactions on*, 34(11):2274–2282. DOI: 10.1109/TPAMI.2012.120. 90

Aharon, M. and Elad, M. (2008). Sparse and redundant modeling of image content using an image-signature-dictionary. *SIAM Journal on Imaging Sciences*, 1(3):228–247. DOI: 10.1137/07070156X. 22, 23, 80, 81

Aharon, M., Elad, M., and Bruckstein, A. (2005). K-SVD: Design of dictionaries for sparse representation. *Proceedings of SPARS*, 5. DOI: 10.1109/TSP.2006.881199. 1, 17, 19, 21, 38, 73, 74

Aharon, M., Elad, M., and Bruckstein, A. (2006). K-svd: An algorithm for designing over-complete dictionaries for sparse representation. *Signal Processing, IEEE Transactions on*, 54(11):4311–4322. DOI: 10.1109/TSP.2006.881199. 77

Alexe, B., Deselaers, T., and Ferrari, V. (2012). Measuring the objectness of image windows. *Pattern Analysis and Machine Intelligence, IEEE Transactions on*, 34(11):2189–2202. DOI: 10.1109/TPAMI.2012.28. 95

Ankerst, M., Breunig, M. M., Kriegel, H.-P., and Sander, J. (1999). Optics: Ordering points to identify the clustering structure. In *ACM Sigmod Record*, volume 28, pages 49–60. ACM. DOI: 10.1145/304181.304187. 6

Arbelaez, P., Maire, M., Fowlkes, C., and Malik, J. (2011). Contour detection and hierarchical image segmentation. *IEEE Trans. Pattern Anal. Mach. Intell.*, 33(5):898–916. DOI: 10.1109/TPAMI.2010.161. 90

Asif, M. S. and Romberg, J. (2013). Sparse recovery of streaming signals using ℓ_1-homotopy. *arXiv preprint arXiv:1306.3331*. DOI: 10.1109/TSP.2014.2328981. 11

Ataee, M., Zayyani, H., Babaie-Zadeh, M., and Jutten, C. (2010). Parametric dictionary learning using steepest descent. In *Acoustics Speech and Signal Processing (ICASSP), 2010 IEEE International Conference on*, pages 1978–1981. IEEE. DOI: 10.1109/ICASSP.2010.5495278. 29, 78

Bach, F. (2009). High-dimensional non-linear variable selection through hierarchical kernel learning. *arXiv preprint arXiv:0909.0844.* 54

Bai, T. and Li, Y. F. (2012). Robust visual tracking with structured sparse representation appearance model. *Pattern Recognition*, 45(6):2390–2404. DOI: 10.1016/j.patcog.2011.12.004. 62, 63, 100, 101

Bao, C., Cai, J.-F., and Ji, H. (2013a). Fast sparsity-based orthogonal dictionary learning for image restoration. In *Computer Vision (ICCV), 2013 IEEE International Conference on*, pages 3384–3391. IEEE. DOI: 10.1109/ICCV.2013.420. 30, 78, 80, 82

Bao, C., Ji, H., Quan, Y., and Shen, Z. (2013b). l0 norm based dictionary learning by proximal methods with global convergence. In *Proceedings of the IEEE Conference on Computer Vision and Pattern Recognition*, pages 3858–3865. DOI: 10.1109/CVPR.2014.493. 28

Bao, C., Wu, Y., Ling, H., and Ji, H. (2012). Real time robust l1 tracker using accelerated proximal gradient approach. In *Computer Vision and Pattern Recognition (CVPR), 2012 IEEE Conference on*, pages 1830–1837. IEEE. DOI: 10.1109/CVPR.2012.6247881. 101

Baraniuk, R. G., Cevher, V., Duarte, M. F., and Hegde, C. (2010). Model-based compressive sensing. *Information Theory, IEEE Transactions on*, 56(4):1982–2001. DOI: 10.1109/TIT.2010.2040894. 12

Barchiesi, D. and Plumbley, M. D. (2011). Dictionary learning of convolved signals. In *Acoustics, Speech and Signal Processing (ICASSP), 2011 IEEE International Conference on*, pages 5812–5815. IEEE. DOI: 10.1109/ICASSP.2011.5947682. 24, 25, 26

Baron, D., Duarte, M. F., Wakin, M. B., Sarvotham, S., and Baraniuk, R. G. (2009). Distributed compressive sensing. *arXiv preprint arXiv:0901.3403.* 45, 46

Barron, A. R., Cohen, A., Dahmen, W., and DeVore, R. A. (2008). Approximation and learning by greedy algorithms. *The Annals of Statistics*, pages 64–94. DOI: 10.1214/009053607000000631. 10

Becker, S., Bobin, J., and Candès, E. J. (2011a). Nesta: a fast and accurate first-order method for sparse recovery. *SIAM Journal on Imaging Sciences*, 4(1):1–39. DOI: 10.1137/090756855. 11

Becker, S. R., Candès, E. J., and Grant, M. C. (2011b). Templates for convex cone problems with applications to sparse signal recovery. *Mathematical Programming Computation*, 3(3):165–218. DOI: 10.1007/s12532-011-0029-5. 11

Belhumeur, P. N., Hespanha, J. P., and Kriegman, D. (1997). Eigenfaces vs. fisherfaces: Recognition using class specific linear projection. *Pattern Analysis and Machine Intelligence, IEEE Transactions on*, 19(7):711–720. DOI: 10.1109/34.598228. 104

Bengio, S., Pereira, F. C., Singer, Y., and Strelow, D. (2009). Group sparse coding. In *NIPS*, volume 22, pages 82–89. 11

Borji, A., Sihite, D. N., and Itti, L. (2013). Quantitative analysis of human-model agreement in visual saliency modeling: A comparative study. *Image Processing, IEEE Transactions on*, 22(1):55–69. DOI: 10.1109/TIP.2012.2210727. 95

Boureau, Y.-L., Bach, F., LeCun, Y., and Ponce, J. (2010). Learning mid-level features for recognition. In *Computer Vision and Pattern Recognition (CVPR), 2010 IEEE Conference on*, pages 2559–2566. IEEE. DOI: 10.1109/CVPR.2010.5539963. 37, 95

Bristow, H., Eriksson, A., and Lucey, S. (2013). Fast convolutional sparse coding. In *Computer Vision and Pattern Recognition (CVPR), 2013 IEEE Conference on*, pages 391–398. IEEE. DOI: 10.1109/CVPR.2013.57. 24, 25

Bruce, N. and Tsotsos, J. (2005). Saliency based on information maximization. In *Advances in Neural Information Processing Systems*, pages 155–162. 96

Bryt, O. and Elad, M. (2008). Compression of facial images using the k-svd algorithm. *Journal of Visual Communication and Image Representation*, 19(4):270–282. DOI: 10.1016/j.jvcir.2008.03.001. 76, 77

Cai, J. F., Candes, E. J., and Shen, Z. (2008). A singular value thresholding algorithm for matrix completion. *preprint*. DOI: 10.1137/080738970. 13, 48

Cai, J.-F., Candès, E. J., and Shen, Z. (2010). A singular value thresholding algorithm for matrix completion. *SIAM Journal on Optimization*, 20(4):1956–1982. DOI: 10.1137/080738970. 14

Candes, E. and Plan, Y. (2009). Matrix completion with noise. *Proceedings of the IEEE*. DOI: 10.1109/JPROC.2009.2035722. 12, 13, 47

Candes, E. and Romberg, J. (2005). l1-magic: Recovery of sparse signals via convex programming. *URL: www. acm. caltech. edu/l1magic/downloads/l1magic. pdf*, 4. 11

Candes, E. J., Romberg, J. K., and Tao, T. (2006). Stable signal recovery from incomplete and inaccurate measurements. *Communications on Pure and Applied Mathematics*, 59(8):1207–1223. DOI: 10.1002/cpa.20124. 10

Cevher, V., Indyk, P., Carin, L., and Baraniuk, R. G. (2010). Sparse signal recovery and acquisition with graphical models. *Signal Processing Magazine, IEEE*, 27(6):92–103. DOI: 10.1109/MSP.2010.938029. 17

Chang, H., Yeung, D.-Y., and Xiong, Y. (2004).   Super-resolution through neighbor embedding.   In *Computer Vision and Pattern Recognition, 2004. CVPR 2004. Proceedings of the 2004 IEEE Computer Society Conference on*, volume 1, pages I–I. IEEE. DOI: 10.1109/CVPR.2004.1315043. 89

Charles, A. S., Olshausen, B. A., and Rozell, C. J. (2011).   Learning sparse codes for hyperspectral imagery.   *Selected Topics in Signal Processing, IEEE Journal of*, 5(5):963–978. DOI: 10.1109/JSTSP.2011.2149497. 80, 82

Chen, S., Cowan, C., and Grant, P. (1991).   Orthogonal least squares learning algorithm for radial basis function networks. *Neural Networks, IEEE Transactions on*, 2(2):302–309. DOI: 10.1109/72.80341. 9, 11

Chen, Y., Nasrabadi, N. M., and Tran, T. D. (2011).   Hyperspectral image classification using dictionary-based sparse representation. *Geoscience and Remote Sensing, IEEE Transactions on*, 49(10):3973–3985. DOI: 10.1109/TGRS.2011.2129595. 90

Chen, Y.-C., Patel, V. M., Phillips, P. J., and Chellappa, R. (2012).   Dictionary-based face recognition from video.   In *Computer Vision–ECCV 2012*, pages 766–779. Springer. DOI: 10.1007/978-3-642-33783-3_55. 94

Chen, Y.-C., Patel, V. M., Pillai, J. K., Chellappa, R., and Phillips, P. J. (2013a).   Dictionary learning from ambiguously labeled data. In *Computer Vision and Pattern Recognition (CVPR), 2013 IEEE Conference on*, pages 353–360. IEEE. DOI: 10.1109/CVPR.2013.52. 42

Chen, Y.-C., Sastry, C. S., Patel, V. M., Phillips, P. J., and Chellappa, R. (2013b).   In-plane rotation and scale invariant clustering using dictionaries. *IEEE Transactions on Image Processing*, 22(6):2166–2180. DOI: 10.1109/TIP.2013.2246178. 44

Chen, Z. and Wu, Y. (2013).   Robust dictionary learning by error source decomposition.   In *Computer Vision (ICCV), 2013 IEEE International Conference on*, pages 2216–2223. IEEE. DOI: 10.1109/ICCV.2013.276. 83

Cheney, E. W. and Cline, A. K. (2004). Topics in sparse approximation. 5

Chi, Y.-T., Ali, M., Rajwade, A., and Ho, J. (2013). Block and group regularized sparse modeling for dictionary learning. In *Computer Vision and Pattern Recognition (CVPR), 2013 IEEE Conference on*, pages 377–382. IEEE. DOI: 10.1109/CVPR.2013.55. 41, 90, 94

Chiang, C.-K., Su, T.-F., Yen, C., and Lai, S.-H. (2013).   Multi-attributed dictionary learning for sparse coding.   In *Computer Vision (ICCV), 2013 IEEE International Conference on*, pages 1137–1144. IEEE. DOI: 10.1109/ICCV.2013.145. 43

Coates, A. and Ng, A. Y. (2011). The importance of encoding versus training with sparse coding and vector quantization. In *Proceedings of the 28th International Conference on Machine Learning (ICML-11)*, pages 921–928. 6, 95

Comaniciu, D., Ramesh, V., and Meer, P. (2003). Kernel-based object tracking. *Pattern Analysis and Machine Intelligence, IEEE Transactions on*, 25(5):564–577. DOI: 10.1109/TPAMI.2003.1195991. 100

Cong, Z., Xiaogang, W., and Wai-Kuen, C. (2011). Background subtraction via robust dictionary learning. *EURASIP Journal on Image and Video Processing*, 2011. DOI: 10.1155/2011/972961. 92

Dabov, K., Foi, A., Katkovnik, V., and Egiazarian, K. (2006). Image denoising with block-matching and 3d filtering. In *Electronic Imaging 2006*, pages 606414–606414. International Society for Optics and Photonics. DOI: 10.1117/12.643267. 80

Dikmen, O. and Févotte, C. (2012). Maximum marginal likelihood estimation for nonnegative dictionary learning in the gamma-poisson model. *Signal Processing, IEEE Transactions on*, 60(10):5163–5175. DOI: 10.1109/TSP.2012.2207117. 67

Ding, X., He, L., and Carin, L. (2011). Bayesian robust principal component analysis. *Image Processing, IEEE Transactions on*, 20(12):3419–3430. DOI: 10.1109/TIP.2011.2156801. 68, 69, 92

Dobigeon, N. and Tourneret, J.-Y. (2010). Bayesian orthogonal component analysis for sparse representation. *Signal Processing, IEEE Transactions on*, 58(5):2675–2685. DOI: 10.1109/TSP.2010.2041594. 65, 66, 75

Dong, W., Li, X., Zhang, D., and Shi, G. (2011a). Sparsity-based image denoising via dictionary learning and structural clustering. In *Computer Vision and Pattern Recognition (CVPR), 2011 IEEE Conference on*, pages 457–464. IEEE. DOI: 10.1109/CVPR.2011.5995478. 79, 80, 81

Dong, W., Zhang, D., and Shi, G. (2011b). Centralized sparse representation for image restoration. In *Computer Vision (ICCV), 2011 IEEE International Conference on*, pages 1259–1266. IEEE. DOI: 10.1109/ICCV.2011.6126377. 80, 88, 89

Dong, W., Zhang, D., Shi, G., and Wu, X. (2011c). Image deblurring and super-resolution by adaptive sparse domain selection and adaptive regularization. *Image Processing, IEEE Transactions on*, 20(7):1838–1857. DOI: 10.1109/TIP.2011.2108306. 88

Donoho, D., Drori, I., Stodden, V., and Tsaig, Y. (2005). Sparselab. software. 11

Donoho, D. L. (2006). Compressed sensing. *Information Theory, IEEE Transactions on*, 52(4):1289–1306. DOI: 10.1109/TIT.2006.871582. 9

Duarte-Carvajalino, J. M. and Sapiro, G. (2009). Learning to sense sparse signals: Simultaneous sensing matrix and sparsifying dictionary optimization. *Image Processing, IEEE Transactions on*, 18(7):1395–1408. DOI: 10.1109/TIP.2009.2022459. 31

Elad, M. and Aharon, M. (2006). Image denoising via sparse and redundant representations over learned dictionaries. *Image Processing, IEEE Transactions on*, 15(12):3736–3745. DOI: 10.1109/TIP.2006.881969. 9

Elad, M., Figueiredo, M. A., and Ma, Y. (2010). On the role of sparse and redundant representations in image processing. *Proceedings of the IEEE*, 98(6):972–982. DOI: 10.1109/JPROC.2009.2037655. 78

Elhamifar, E., Sapiro, G., and Vidal, R. (2012). See all by looking at a few: Sparse modeling for finding representative objects. In *Computer Vision and Pattern Recognition (CVPR), 2012 IEEE Conference on*, pages 1600–1607. IEEE. DOI: 10.1109/CVPR.2012.6247852. 32

Engan, K., Aase, S. O., and Hakon Husoy, J. (1999a). Method of optimal directions for frame design. In *Acoustics, Speech, and Signal Processing, 1999. Proceedings., 1999 IEEE International Conference on*, volume 5, pages 2443–2446. IEEE. DOI: 10.1109/ICASSP.1999.760624. 19

Engan, K., Aase, S. O., and Husoy, J. (1999b). Frame based signal compression using method of optimal directions (mod). In *Circuits and Systems, 1999. ISCAS'99. Proceedings of the 1999 IEEE International Symposium on*, volume 4, pages 1–4. IEEE. DOI: 10.1109/IS-CAS.1999.779928. 19

Engan, K., Aase, S. O., and Husøy, J. H. (2000). Multi-frame compression: Theory and design. *Signal Processing*, 80(10):2121–2140. DOI: 10.1016/S0165-1684(00)00072-4. 5

Esser, E., Moller, M., Osher, S., Sapiro, G., and Xin, J. (2012). A convex model for nonnegative matrix factorization and dimensionality reduction on physical space. *Image Processing, IEEE Transactions on*, 21(7):3239–3252. DOI: 10.1109/TIP.2012.2190081. 30, 94

Ester, M., Kriegel, H.-P., Sander, J., and Xu, X. (1996). A density-based algorithm for discovering clusters in large spatial databases with noise. In *Kdd*, volume 96, pages 226–231. 6

Etemad, K. and Chellappa, R. (1998). Separability-based multiscale basis selection and feature extraction for signal and image classification. *Image Processing, IEEE Transactions on*, 7(10):1453–1465. DOI: 10.1109/83.718485. 56

Farsiu, S., Robinson, M. D., Elad, M., and Milanfar, P. (2004). Fast and robust multi-frame super resolution. *Image processing, IEEE Transactions on*, 13(10):1327–1344. DOI: 10.1109/TIP.2004.834669. 87

Feng, J., Ni, B., Tian, Q., and Yan, S. (2011a). Geometric lp-norm feature pooling for image classification. In *Computer Vision and Pattern Recognition (CVPR), 2011 IEEE Conference on*, pages 2609–2704. IEEE. DOI: 10.1109/CVPR.2011.5995370. 95

Feng, J., Song, L., Yang, X., and Zhang, W. (2011b). Learning dictionary via subspace segmentation for sparse representation. In *Image Processing (ICIP), 2011 18th IEEE International Conference on*, pages 1245–1248. IEEE. DOI: 10.1109/ICIP.2011.6115658. 43, 44, 80

Figueiredo, M. A., Nowak, R. D., and Wright, S. J. (2007). Gradient projection for sparse reconstruction: Application to compressed sensing and other inverse problems. *Selected Topics in Signal Processing, IEEE Journal of*, 1(4):586–597. DOI: 10.1109/JSTSP.2007.910281. 104

Friedman, J., Hastie, T., and Tibshirani, R. (2010). A note on the group lasso and a sparse group lasso. *arXiv preprint arXiv:1001.0736*. 12

Gao, J., Guo, Y., and Yin, M. (2013). Restricted boltzmann machine approach to couple dictionary training for image super-resolution. In *IEEE International Conference on Image Processing*, pages 499–503. 89

Gao, S., Tsang, I. W., Chia, L.-T., and Zhao, P. (2010). Local features are not lonely–laplacian sparse coding for image classification. In *Computer Vision and Pattern Recognition (CVPR), 2010 IEEE Conference on*, pages 3555–3561. IEEE. DOI: 10.1109/ICIP.2013.6738103. 31

Gersho, A. and Gray, R. M. (1992). *Vector Quantization and Signal Compression*. Springer. DOI: 10.1007/978-1-4615-3626-0. 6

Glasner, D., Bagon, S., and Irani, M. (2009). Super-resolution from a single image. In *Computer Vision, 2009 IEEE 12th International Conference on*, pages 349–356. IEEE. DOI: 10.1109/ICCV.2009.5459271. 87

Goodfellow, I., Courville, A., and Bengio, Y. (2012). Large-scale feature learning with spike-and-slab sparse coding. *arXiv preprint arXiv:1206.6407*. 15, 16, 17

Guha, T. and Ward, R. K. (2012). Learning sparse representations for human action recognition. *Pattern Analysis and Machine Intelligence, IEEE Transactions on*, 34(8):1576–1588. DOI: 10.1109/TPAMI.2011.253. 9

Gunturk, B. K., Altunbasak, Y., and Mersereau, R. M. (2002). Color plane interpolation using alternating projections. *Image Processing, IEEE Transactions on*, 11(9):997–1013. DOI: 10.1109/TIP.2002.801121. 84

Guo, H., Jiang, Z., and Davis, L. S. (2013). Discriminative dictionary learning with pairwise constraints. In *Computer Vision–ACCV 2012*, pages 328–342. Springer. DOI: 10.1007/978-3-642-37331-2_25. 36

Hale, E. T., Yin, W., and Zhang, Y. (2008). Fixed-point continuation for \ell_1-minimization: Methodology and convergence. *SIAM Journal on Optimization*, 19(3):1107–1130. DOI: 10.1137/070698920. 48

Harandi, M. T., Sanderson, C., Hartley, R., and Lovell, B. C. (2012). Sparse coding and dictionary learning for symmetric positive definite matrices: A kernel approach. In *Computer Vision–ECCV 2012*, pages 216–229. Springer. DOI: 10.1007/978-3-642-33709-3_16. 27

Hardie, R. C., Barnard, K. J., and Armstrong, E. E. (1997). Joint map registration and high-resolution image estimation using a sequence of undersampled images. *Image Processing, IEEE Transactions on*, 6(12):1621–1633. DOI: 10.1109/83.650116. 87

Hawe, S., Seibert, M., and Kleinsteuber, M. (2013). Separable dictionary learning. In *Computer Vision and Pattern Recognition (CVPR), 2013 IEEE Conference on*, pages 438–445. IEEE. DOI: 10.1109/CVPR.2013.63. 50, 51, 78, 80

He, L., Qi, H., and Zaretzki, R. (2013). Beta process joint dictionary learning for coupled feature spaces with application to single image super-resolution. In *Computer Vision and Pattern Recognition (CVPR), 2013 IEEE Conference on*, pages 345–352. IEEE. DOI: 10.1109/CVPR.2013.51. 68, 69, 70, 88

Hou, X. and Zhang, L. (2009). Dynamic visual attention: Searching for coding length increments. In *Advances in Neural Information Processing Systems*, pages 681–688. 96

Huang, D.-A. and Wang, Y.-C. F. (2013). Coupled dictionary and feature space learning with applications to cross-domain image synthesis and recognition. In *Computer Vision (ICCV), 2013 IEEE International Conference on*, pages 2496–2503. IEEE. DOI: 10.1109/ICCV.2013.310. 52, 88

Huang, J., Zhang, T., and Metaxas, D. (2011). Learning with structured sparsity. *The Journal of Machine Learning Research*, 12:3371–3412. 12

Irie, G., Liu, D., Li, Z., and Chang, S.-F. (2013). A bayesian approach to multimodal visual dictionary learning. In *Computer Vision and Pattern Recognition (CVPR), 2013 IEEE Conference on*, pages 329–336. IEEE. DOI: 10.1109/CVPR.2013.49. 70, 71

Itti, L., Koch, C., and Niebur, E. (1998). A model of saliency-based visual attention for rapid scene analysis. *IEEE Transactions on Pattern Analysis and Machine Intelligence*, 20(11):1254–1259. DOI: 10.1109/34.730558. 95, 96

Jafari, M. G. and Plumbley, M. D. (2011). Fast dictionary learning for sparse representations of speech signals. *Selected Topics in Signal Processing, IEEE Journal of*, 5(5):1025–1031. DOI: 10.1109/JSTSP.2011.2157892. 28, 77

Jenatton, R., Mairal, J., Bach, F. R., and Obozinski, G. R. (2010). Proximal methods for sparse hierarchical dictionary learning. In *Proceedings of the 27th International Conference on Machine Learning (ICML-10)*, pages 487–494. 54, 55

Jenatton, R., Mairal, J., Obozinski, G., and Bach, F. (2011). Proximal methods for hierarchical sparse coding. *The Journal of Machine Learning Research*, 12:2297–2334. 54

Ji, S., Xue, Y., and Carin, L. (2008). Bayesian compressive sensing. *Signal Processing, IEEE Transactions on*, 56(6):2346–2356. DOI: 10.1109/TSP.2007.914345. 14

Jiang, M., Song, M., and Zhao, Q. (2013). Leveraging human fixations in sparse coding: Learning a discriminative dictionary for saliency prediction. In *Systems, Man, and Cybernetics (SMC), 2013 IEEE International Conference on*, pages 2126–2133. IEEE. DOI: 10.1109/SMC.2013.364. 97

Jiang, Z., Lin, Z., and Davis, L. S. (2011). Learning a discriminative dictionary for sparse coding via label consistent k-svd. In *Computer Vision and Pattern Recognition (CVPR), 2011 IEEE Conference on*, pages 1697–1704. IEEE. DOI: 10.1109/CVPR.2011.5995354. 35, 37, 38, 94, 97

Jiang, Z., Zhang, G., and Davis, L. S. (2012). Submodular dictionary learning for sparse coding. In *Computer Vision and Pattern Recognition (CVPR), 2012 IEEE Conference on*, pages 3418–3425. IEEE. DOI: 10.1109/CVPR.2012.6248082. 29, 76

Kamilov, U., Rangan, S., Unser, M., and Fletcher, A. K. (2012). Approximate message passing with consistent parameter estimation and applications to sparse learning. In *Advances in Neural Information Processing Systems*, pages 2438–2446. DOI: 10.1109/TIT.2014.2309005. 11

Kang, L.-W., Lin, C.-W., and Fu, Y.-H. (2012). Automatic single-image-based rain streaks removal via image decomposition. *Image Processing, IEEE Transactions on*, 21(4):1742–1755. DOI: 10.1109/TIP.2011.2179057. 84

Kasiviswanathan, S. P., Melville, P., Banerjee, A., and Sindhwani, V. (2011). Emerging topic detection using dictionary learning. In *Proceedings of the 20th ACM international conference on Information and knowledge management*, pages 745–754. ACM. DOI: 10.1145/2063576.2063686. 98

Kaufman, L. and Rousseeuw, P. (1987). *Clustering by Means of Medoids*. North-Holland. 32

Kavukcuoglu, K., Sermanet, P., Boureau, Y.-L., Gregor, K., Mathieu, M., and LeCun, Y. (2010). Learning convolutional feature hierarchies for visual recognition. In *NIPS*, volume 1, page 5. 26

Kim, S. and Xing, E. P. (2010). Tree-guided group lasso for multi-task regression with structured sparsity. In *Proceedings of the 27th International Conference on Machine Learning (ICML-10)*, pages 543–550. 54

Koh, K., Kim, S.-J., and Boyd, S. P. (2007). An interior-point method for large-scale l1-regularized logistic regression. *Journal of Machine Learning Research*, 8(8):1519–1555. 11

Kolter, J. Z., Batra, S., and Ng, A. Y. (2010). Energy disaggregation via discriminative sparse coding. In *NIPS*, pages 1153–1161. 93, 94

Kong, S. and Wang, D. (2012). A dictionary learning approach for classification: separating the particularity and the commonality. In *Computer Vision–ECCV 2012*, pages 186–199. Springer. DOI: 10.1007/978-3-642-33718-5_14. 49

Kong, X., Li, K., Cao, J., Yang, Q., and Wenyin, L. (2013). Hep-2 cell pattern classification with discriminative dictionary learning. *Pattern Recognition*. DOI: 10.1016/j.patcog.2013.09.025. 40, 94

Kreutz-Delgado, K., Murray, J. F., Rao, B. D., Engan, K., Lee, T.-W., and Sejnowski, T. J. (2003). Dictionary learning algorithms for sparse representation. *Neural Computation*, 15(2):349–396. DOI: 10.1162/089976603762552951. 5

Labusch, K., Barth, E., and Martinetz, T. (2009). Sparse coding neural gas: learning of overcomplete data representations. *Neurocomputing*, 72(7):1547–1555. DOI: 10.1016/j.neucom.2008.11.027. 28

Lang, C., Liu, G., Yu, J., and Yan, S. (2012). Saliency detection by multitask sparsity pursuit. *Image Processing, IEEE Transactions on*, 21(3):1327–1338. DOI: 10.1109/TIP.2011.2169274. 95

Li, H., Shen, C., and Shi, Q. (2011a). Real-time visual tracking using compressive sensing. In *Computer Vision and Pattern Recognition (CVPR), 2011 IEEE Conference on*, pages 1305–1312. IEEE. DOI: 10.1109/CVPR.2011.5995483. 101

Li, L., Silva, J., Zhou, M., and Carin, L. (2012). Online bayesian dictionary learning for large datasets. In *Acoustics, Speech and Signal Processing (ICASSP), 2012 IEEE International Conference on*, pages 2157–2160. IEEE. DOI: 10.1109/ICASSP.2012.6288339. 65, 82

Li, L., Zhou, M., Wang, E., and Carin, L. (2011b). Joint dictionary learning and topic modeling for image clustering. In *ICASSP*, pages 2168–2171. DOI: 10.1109/ICASSP.2011.5946757. 65

Li, P., Wang, Q., Zuo, W., and Zhang, L. (2013). Log-euclidean kernels for sparse representation and dictionary learning. In *Computer Vision (ICCV), 2013 IEEE International Conference on*, pages 1601–1608. IEEE. DOI: 10.1109/ICCV.2013.202. 27

Li, Q., Zhou, Y., and Yang, J. (2011c). Saliency based image segmentation. In *Multimedia Technology (ICMT), 2011 International Conference on*, pages 5068–5071. IEEE. DOI: 10.1109/ICMT.2011.6002178. 95

Li, S. and Fang, L. (2010). An efficient learned dictionary and its application to non-local denoising. In *Image Processing (ICIP), 2010 17th IEEE International Conference on*, pages 1945–1948. IEEE. DOI: 10.1109/ICIP.2010.5652718. 43, 44, 80

Li, X. and Haupt, J. (2014). Identifying outliers in large matrices via randomized adaptive compressive sampling. *arXiv preprint arXiv:1407.0312*. DOI: 10.1109/ICIP.2010.5652718. 96, 97

Lian, X.-C., Li, Z., Wang, C., Lu, B.-L., and Zhang, L. (2010). Probabilistic models for supervised dictionary learning. In *Computer Vision and Pattern Recognition (CVPR), 2010 IEEE Conference on*, pages 2305–2312. IEEE. DOI: 10.1109/CVPR.2010.5539915. 67

Ling, H., Bai, L., Blasch, E., and Mei, X. (2010). Robust infrared vehicle tracking across target pose change using l 1 regularization. In *Information Fusion (FUSION), 2010 13th Conference on*, pages 1–8. IEEE. DOI: 10.1109/ICIF.2010.5711902. 99

Liu, B., Huang, J., Yang, L., and Kulikowsk, C. (2011a). Robust tracking using local sparse appearance model and k-selection. In *Computer Vision and Pattern Recognition (CVPR), 2011 IEEE Conference on*, pages 1313–1320. IEEE. DOI: 10.1109/CVPR.2011.5995730. 9

Liu, B., Yang, L., Huang, J., Meer, P., Gong, L., and Kulikowski, C. (2010). Robust and fast collaborative tracking with two stage sparse optimization. In *Computer Vision–ECCV 2010*, pages 624–637. Springer. DOI: 10.1007/978-3-642-15561-1_45. 101

Liu, D. and Boufounos, P. T. (2012). Dictionary learning based pan-sharpening. In *Acoustics, Speech and Signal Processing (ICASSP), 2012 IEEE International Conference on*, pages 2397–2400. IEEE. DOI: 10.1109/ICASSP.2012.6288398. 90

Liu, H., Sun, F., and Gao, M. (2011b). Visual tracking using iterative sparse approximation. In *Advances in Neural Networks–ISNN 2011*, pages 207–214. Springer. DOI: 10.1007/978-3-642-21090-7_25. 101

Liu, J., Chen, S., and Tan, X. (2008). Fractional order singular value decomposition representation for face recognition. *Pattern Recognition*, 41(1):378–395. DOI: 10.1016/j.patcog.2007.03.027. 46

Liu, J., Tai, X.-C., Huang, H., and Huan, Z. (2013a). A weighted dictionary learning model for denoising images corrupted by mixed noise. *Image Processing, IEEE Transactions on*, 22(3):1108–1120. DOI: 10.1109/TIP.2012.2227766. 79, 80

Liu, Q., Wang, S., Ying, L., Peng, X., Zhu, Y., and Liang, D. (2013b). Adaptive dictionary learning in sparse gradient domain for image recovery. *Image Processing, IEEE Transactions on*, 22(12):4652–4663. DOI: 10.1109/TIP.2013.2277798. 86

Liu, X., Song, M., Tao, D., Zhou, X., Chen, C., and Bu, J. (2013c). Semi-supervised coupled dictionary learning for person re-identification. In *Proceedings of the IEEE Conference on Computer Vision and Pattern Recognition*, pages 3550–3557. DOI: 10.1109/CVPR.2014.454. 52

Lu, C., Shi, J., and Jia, J. (2013). Online robust dictionary learning. In *Computer Vision and Pattern Recognition (CVPR), 2013 IEEE Conference on*, pages 415–422. IEEE. DOI: 10.1109/CVPR.2013.60. 61, 83, 92

Luo, J., Wang, W., and Qi, H. (2013). Group sparsity and geometry constrained dictionary learning for action recognition from depth maps. In *Computer Vision (ICCV), 2013 IEEE International Conference on*, pages 1809–1816. IEEE. DOI: 10.1109/ICCV.2013.227. 41, 94

Ma, L., Wang, C., Xiao, B., and Zhou, W. (2012). Sparse representation for face recognition based on discriminative low-rank dictionary learning. In *Computer Vision and Pattern Recognition (CVPR), 2012 IEEE Conference on*, pages 2586–2593. IEEE. DOI: 10.1109/CVPR.2012.6247977. 44

MacQueen, J. et al. (1967). Some methods for classification and analysis of multivariate observations. In *Proceedings of the Fifth Berkeley Symposium on Mathematical Statistics and Probability*, volume 1, pages 281–297. California, USA. 5

Mahmoudi, M. and Sapiro, G. (2005). Fast image and video denoising via nonlocal means of similar neighborhoods. *Signal Processing Letters, IEEE*, 12(12):839–842. DOI: 10.1109/LSP.2005.859509. 79

Mailhé, B., Lesage, S., Gribonval, R., Bimbot, F., Vandergheynst, P., et al. (2008). Shift-invariant dictionary learning for sparse representations: extending k-svd. In *16th European Signal Processing Conference (EUSIPCO'08)*. 23, 92

Mairal, J., Bach, F., and Ponce, J. (2012). Task-driven dictionary learning. *Pattern Analysis and Machine Intelligence, IEEE Transactions on*, 34(4):791–804. DOI: 10.1109/TPAMI.2011.156. 40

Mairal, J., Bach, F., Ponce, J., and Sapiro, G. (2009a). Online dictionary learning for sparse coding. In *Proceedings of the 26th Annual International Conference on Machine Learning*, pages 689–696. ACM. DOI: 10.1145/1553374.1553463. 59, 60, 61, 82, 83

Mairal, J., Bach, F., Ponce, J., Sapiro, G., and Zisserman, A. (2008a). Discriminative learned dictionaries for local image analysis. In *Computer Vision and Pattern Recognition, 2008. CVPR 2008. IEEE Conference on*, pages 1–8. IEEE. DOI: 10.1109/CVPR.2008.4587652. 39, 40, 90, 94

Mairal, J., Bach, F., Ponce, J., Sapiro, G., and Zisserman, A. (2009b). Non-local sparse models for image restoration. In *Computer Vision, 2009 IEEE 12th International Conference on*, pages 2272–2279. IEEE. DOI: 10.1109/ICCV.2009.5459452. 31, 80, 83, 84, 85, 86

Mairal, J., Elad, M., and Sapiro, G. (2008b). Sparse representation for color image restoration. *Image Processing, IEEE Transactions on*, 17(1):53–69. DOI: 10.1109/TIP.2007.911828. 84, 85, 86

Mairal, J., Ponce, J., Sapiro, G., Zisserman, A., and Bach, F. R. (2009c). Supervised dictionary learning. In *Advances in Neural Information Processing Systems*, pages 1033–1040. 37, 90, 94, 105

Mallat, S. G. and Zhang, Z. (1993). Matching pursuits with time-frequency dictionaries. *Signal Processing, IEEE Transactions on*, 41(12):3397–3415. DOI: 10.1109/78.258082. 10

Mazaheri, J. A., Guillemot, C., and Labit, C. (2013). Learning a tree-structured dictionary for efficient image representation with adaptive sparse coding. In *Acoustics, Speech and Signal Processing (ICASSP), 2013 IEEE International Conference on*, pages 1320–1324. IEEE. DOI: 10.1109/ICASSP.2013.6637865. 57, 76

Mazhar, R. and Gader, P. D. (2008). Ek-svd: Optimized dictionary design for sparse representations. In *Pattern Recognition, 2008. ICPR 2008. 19th International Conference on*, pages 1–4. IEEE. DOI: 10.1109/ICPR.2008.4761362. 29

Mei, X. and Ling, H. (2009). Robust visual tracking using &# x2113; 1 minimization. In *Computer Vision, 2009 IEEE 12th International Conference on*, pages 1436–1443. IEEE. DOI: 10.1109/ICCV.2009.5459292. 98, 99, 100, 101

Mei, X., Ling, H., Wu, Y., Blasch, E., and Bai, L. (2011). Minimum error bounded efficient ℓ_1 tracker with occlusion detection. In *Computer Vision and Pattern Recognition (CVPR), 2011 IEEE Conference on*, pages 1257–1264. IEEE. DOI: 10.1109/CVPR.2011.5995421. 102

Meka, R., Jain, P., and Dhillon, I. S. (2009). Matrix completion from power-law distributed samples. In *Advances in neural information processing systems*, pages 1258–1266. 14

Mysore, G. J. (2012). A block sparsity approach to multiple dictionary learning for audio modeling. In *International Conference on Machine Learning (ICML) Workshop on Sparsity, Dictionaries, and Projections in Machine Learning and Signal Processing*. 78

Nagesh, P. and Li, B. (2009). A compressive sensing approach for expression-invariant face recognition. In *Computer Vision and Pattern Recognition, 2009. CVPR 2009. IEEE Conference on*, pages 1518–1525. IEEE. DOI: 10.1109/CVPRW.2009.5206657. 45, 46

Najman, L. and Schmitt, M. (1994). Watershed of a continuous function. *Signal Processing*, 38(1):99–112. DOI: 10.1016/0165-1684(94)90059-0. 90

Nasrabadi, N. M. and King, R. A. (1988). Image coding using vector quantization: A review. *Communications, IEEE Transactions on*, 36(8):957–971. DOI: 10.1109/26.3776. 73

Nasrollahi, K. and Moeslund, T. B. (2014). Super-resolution: A comprehensive survey. *Machine Vision & Applications*. DOI: 10.1007/s00138-014-0623-4. 86

Needell, D. and Vershynin, R. (2009). Uniform uncertainty principle and signal recovery via regularized orthogonal matching pursuit. *Foundations of Computational Mathematics*, 9(3):317–334. DOI: 10.1007/s10208-008-9031-3. 11

Nguyen, H., Patel, V. M., Nasrabadi, N. M., and Chellappa, R. (2012). Kernel dictionary learning. In *Acoustics, Speech and Signal Processing (ICASSP), 2012 IEEE International Conference on*, pages 2021–2024. IEEE. DOI: 10.1109/ICASSP.2012.6288305. 27

Ni, J., Qiu, Q., and Chellappa, R. (2013). Subspace interpolation via dictionary learning for unsupervised domain adaptation. In *Computer Vision and Pattern Recognition (CVPR), 2013 IEEE Conference on*, pages 692–699. IEEE. DOI: 10.1109/CVPR.2013.95. 53

Oja, E. (1982). Simplified neuron model as a principal component analyzer. *Journal of Mathematical Biology*, 15(3):267–273. DOI: 10.1007/BF00275687. 28

Ophir, B., Lustig, M., and Elad, M. (2011). Multi-scale dictionary learning using wavelets. *Selected Topics in Signal Processing, IEEE Journal of*, 5(5):1014–1024. DOI: 10.1109/JSTSP.2011.2155032. 56, 57, 76

Osher, S. and Paragios, N. (2003). *Geometric Level Set Methods in Imaging, Vision, and Graphics*. Springer. DOI: 10.1007/b97541. 90

Paliy, D., Katkovnik, V., Bilcu, R., Alenius, S., and Egiazarian, K. (2007). Spatially adaptive color filter array interpolation for noiseless and noisy data. *International Journal of Imaging Systems and Technology*, 17(3):105–122. DOI: 10.1002/ima.20109. 84

Patel, V. M., Easley, G. R., Chellappa, R., and Nasrabadi, N. M. (2014). Separated component-based restoration of speckled sar images. *Geoscience and Remote Sensing, IEEE Transactions on*, 52(2):1019–1029. DOI: 10.1109/TGRS.2013.2246794. 82

Patel, V. M., Nasrabadi, N. M., and Chellappa, R. (2011). Sparsity-motivated automatic target recognition. *Applied Optics*, 50(10):1425–1433. DOI: 10.1364/AO.50.001425. 94

Patel, V. M., Wu, T., Biswas, S., Phillips, P. J., and Chellappa, R. (2012). Dictionary-based face recognition under variable lighting and pose. *Information Forensics and Security, IEEE Transactions on*, 7(3):954–965. DOI: 10.1109/TIFS.2012.2189205. 94

Pati, Y. C., Rezaiifar, R., and Krishnaprasad, P. (1993). Orthogonal matching pursuit: Recursive function approximation with applications to wavelet decomposition. In *Signals, Systems and Computers, 1993. 1993 Conference Record of The Twenty-Seventh Asilomar Conference on*, pages 40–44. IEEE. DOI: 10.1109/ACSSC.1993.342465. 1, 10, 104

Peleg, T., Eldar, Y. C., and Elad, M. (2012). Exploiting statistical dependencies in sparse representations for signal recovery. *Signal Processing, IEEE Transactions on*, 60(5):2286–2303. DOI: 10.1109/TSP.2012.2188520. 71

Peng, Y., Meng, D., Xu, Z., Gao, C., Yang, Y., and Zhang, B. (2013). Decomposable nonlocal tensor dictionary learning for multispectral image denoising. In *Proceedings of the IEEE Conference on Computer Vision and Pattern Recognition*, pages 2949–2956. DOI: 10.1109/CVPR.2014.377. 32, 80, 82

Peyré, G. (2011). A review of adaptive image representations. *Selected Topics in Signal Processing, IEEE Journal of*, 5(5):896–911. DOI: 10.1109/JSTSP.2011.2120592. 79

Pham, D.-S. and Venkatesh, S. (2008). Joint learning and dictionary construction for pattern recognition. In *Computer Vision and Pattern Recognition, 2008. CVPR 2008. IEEE Conference on*, pages 1–8. IEEE. DOI: 10.1109/CVPR.2008.4587408. 32, 34, 37, 94, 105, 107

Porikli, F., Tuzel, O., and Meer, P. (2006). Covariance tracking using model update based on lie algebra. In *Computer Vision and Pattern Recognition, 2006 IEEE Computer Society Conference on*, volume 1, pages 728–735. IEEE. DOI: 10.1109/CVPR.2006.94. 100

Qiu, Q., Jiang, Z., and Chellappa, R. (2011). Sparse dictionary-based representation and recognition of action attributes. In *Computer Vision (ICCV), 2011 IEEE International Conference on*, pages 707–714. IEEE. DOI: 10.1109/ICCV.2011.6126307. 41, 94

Qiu, Q., Patel, V. M., and Chellappa, R. (2014). Information-theoretic dictionary learning for image classification. *IEEE Transactions on Pattern Analysis & Machine Intelligence*, (11):2173–2184. DOI: 10.1109/TPAMI.2014.2316824. 41

Qiu, Q., Patel, V. M., Turaga, P., and Chellappa, R. (2012). Domain adaptive dictionary learning. In *Computer Vision–ECCV 2012*, pages 631–645. Springer. DOI: 10.1007/978-3-642-33765-9_45. 53

Rakotomamonjy, A. (2013). Direct optimization of the dictionary learning problem. *Signal Processing, IEEE Transactions on*, 61(22):5495–5506. DOI: 10.1109/TSP.2013.2278158. 29

Ramirez, I., Sprechmann, P., and Sapiro, G. (2010). Classification and clustering via dictionary learning with structured incoherence and shared features. In *Computer Vision and Pattern Recognition (CVPR), 2010 IEEE Conference on*, pages 3501–3508. IEEE. DOI: 10.1109/CVPR.2010.5539964. 40, 41

Rasmussen, C. E. (1999). The infinite gaussian mixture model. In *NIPS*, volume 12, pages 554–560. 6

Rother, C., Kolmogorov, V., and Blake, A. (2004). Grabcut: Interactive foreground extraction using iterated graph cuts. In *ACM Transactions on Graphics (TOG)*, volume 23, pages 309–314. ACM. DOI: 10.1145/1015706.1015720. 90

Rubinstein, R., Zibulevsky, M., and Elad, M. (2010). Double sparsity: Learning sparse dictionaries for sparse signal approximation. *Signal Processing, IEEE Transactions on*, 58(3):1553–1564. DOI: 10.1109/TSP.2009.2036477. 30, 78, 80

Sastry, S. S. and Ma, Y. (2010). Fast ℓ_1-minimization algorithms for robust face recognition. In *Proceedings of the International Conference on Image Processing in*, page 1. DOI: 10.1109/TIP.2013.2262292. 11

Sharma, G., Jurie, F., and Schmid, C. (2012). Discriminative spatial saliency for image classification. In *Computer Vision and Pattern Recognition (CVPR), 2012 IEEE Conference on*, pages 3506–3513. IEEE. DOI: 10.1109/CVPR.2012.6248093. 95

Shekhar, S., Patel, V. M., Nasrabadi, N. M., and Chellappa, R. (2014). Joint sparse representation for robust multimodal biometrics recognition. *Pattern Analysis and Machine Intelligence, IEEE Transactions on*, 36(1):113–126. DOI: 10.1109/TPAMI.2013.109. 53

Shen, L., Wang, S., Sun, G., Jiang, S., and Huang, Q. (2013). Multi-level discriminative dictionary learning towards hierarchical visual categorization. In *Computer Vision and Pattern Recognition (CVPR), 2013 IEEE Conference on*, pages 383–390. IEEE. DOI: 10.1109/CVPR.2013.56. 57

Shen, X. and Wu, Y. (2012). A unified approach to salient object detection via low rank matrix recovery. In *Computer Vision and Pattern Recognition (CVPR), 2012 IEEE Conference on*, pages 853–860. IEEE. DOI: 10.1109/CVPR.2012.6247758. 95

Shi, J., Ren, X., Dai, G., Wang, J., and Zhang, Z. (2011). A non-convex relaxation approach to sparse dictionary learning. In *Computer Vision and Pattern Recognition (CVPR), 2011 IEEE Conference on*, pages 1809–1816. IEEE. DOI: 10.1109/CVPR.2011.5995592. 78, 80, 82

Sigg, C. D., Dikk, T., and Buhmann, J. M. (2010). Speech enhancement with sparse coding in learned dictionaries. In *Acoustics Speech and Signal Processing (ICASSP), 2010 IEEE International Conference on*, pages 4758–4761. IEEE. DOI: 10.1109/ICASSP.2010.5495157. 85

Sivalingam, R., Boley, D., Morellas, V., and Papanikolopoulos, N. (2011). Positive definite dictionary learning for region covariances. In *Computer Vision (ICCV), 2011 IEEE International Conference on*, pages 1013–1019. IEEE. DOI: 10.1109/ICCV.2011.6126346. 30

Siyahjani, F. and Doretto, G. (2013). Learning a context aware dictionary for sparse representation. In *Computer Vision–ACCV 2012*, pages 228–241. Springer. DOI: 10.1007/978-3-642-37444-9_18. 43

Skretting, K. and Engan, K. (2010). Recursive least squares dictionary learning algorithm. *Signal Processing, IEEE Transactions on*, 58(4):2121–2130. DOI: 10.1109/TSP.2010.2040671. 59, 60, 61

Skretting, K. and Engan, K. (2011). Image compression using learned dictionaries by rls-dla and compared with k-svd. In *Acoustics, Speech and Signal Processing (ICASSP), 2011 IEEE International Conference on*, pages 1517–1520. IEEE. DOI: 10.1109/ICASSP.2011.5946782. 59, 61, 75

Smola, A. J. and Schökopf, B. (2000). Sparse greedy matrix approximation for machine learning. In *Proceedings of the Seventeenth International Conference on Machine Learning*, pages 911–918. Morgan Kaufmann Publishers Inc. 30

Sprechmann, P. and Sapiro, G. (2010). Dictionary learning and sparse coding for unsupervised clustering. In *Acoustics Speech and Signal Processing (ICASSP), 2010 IEEE International Conference on*, pages 2042–2045. IEEE. DOI: 10.1109/ICASSP.2010.5494985. 44

Stauffer, C. and Grimson, W. E. L. (1999). Adaptive background mixture models for real-time tracking. In *Computer Vision and Pattern Recognition, 1999. IEEE Computer Society Conference on.*, volume 2. IEEE. DOI: 10.1109/CVPR.1999.784637. 91

Szabó, Z., Póczos, B., and Lorincz, A. (2011). Online group-structured dictionary learning. In *Computer Vision and Pattern Recognition (CVPR), 2011 IEEE Conference on*, pages 2865–2872. IEEE. DOI: 10.1109/CVPR.2011.5995712. 62, 83, 84

Szekely, G. J. and Rizzo, M. L. (2005). Hierarchical clustering via joint between-within distances: Extending ward's minimum variance method. *Journal of Classification*, 22(2):151–183. DOI: 10.1007/s00357-005-0012-9. 6

Taheri, S., Patel, V. M., and Chellappa, R. (2013). Component-based recognition of facesand facial expressions. *Affective Computing, IEEE Transactions on*, 4(4):360–371. DOI: 10.1109/T-AFFC.2013.28. 48

Tosic, I., Shroff, S. A., and Berkner, K. (2013). Dictionary learning for incoherent sampling with application to plenoptic imaging. In *Acoustics, Speech and Signal Processing (ICASSP), 2013 IEEE International Conference on*, pages 1821–1825. IEEE. DOI: 10.1109/ICASSP.2013.6637967. 90

Tropp, J. A. and Gilbert, A. C. (2007). Signal recovery from random measurements via orthogonal matching pursuit. *Information Theory, IEEE Transactions on*, 53(12):4655–4666. DOI: 10.1109/TIT.2007.909108. 1, 10, 104

Turk, M. and Pentland, A. (1991). Eigenfaces for recognition. *Journal of Cognitive Neuroscience*, 3(1):71–86. DOI: 10.1162/jocn.1991.3.1.71. 104

Tzimiropoulos, G., Zafeiriou, S., and Pantic, M. (2011). Sparse representations of image gradient orientations for visual recognition and tracking. In *Computer Vision and Pattern Recognition Workshops (CVPRW), 2011 IEEE Computer Society Conference on*, pages 26–33. IEEE. DOI: 10.1109/CVPRW.2011.5981809. 99

Van Nguyen, H., Patel, V. M., Nasrabadi, N. M., and Chellappa, R. (2013). Design of nonlinear kernel dictionaries for object recognition. *Image Processing, IEEE Transactions on*, 22(12):5123–5135. DOI: 10.1109/TIP.2013.2282078. 27

Van Ouwerkerk, J. (2006). Image super-resolution survey. *Image and Vision Computing*, 24(10):1039–1052. DOI: 10.1016/j.imavis.2006.02.026. 87

Vicente, S., Kolmogorov, V., and Rother, C. (2008). Graph cut based image segmentation with connectivity priors. In *Computer Vision and Pattern Recognition, 2008. CVPR 2008. IEEE Conference on*, pages 1–8. IEEE. DOI: 10.1109/CVPR.2008.4587440. 90

Wang, J., Yang, J., Yu, K., Lv, F., Huang, T., and Gong, Y. (2010). Locality-constrained linear coding for image classification. In *Computer Vision and Pattern Recognition (CVPR), 2010 IEEE Conference on*, pages 3360–3367. IEEE. DOI: 10.1109/CVPR.2010.5540018. 38

Wang, N., Wang, J., and Yeung, D.-Y. (2013a). Online robust non-negative dictionary learning for visual tracking. In *Computer Vision (ICCV), 2013 IEEE International Conference on*, pages 657–664. IEEE. DOI: 10.1109/ICCV.2013.87. 62

Wang, S., Liu, Q., Xia, Y., Dong, P., Luo, J., Huang, Q., and Feng, D. D. (2013b). Dictionary learning based impulse noise removal via l1–l1 minimization. *Signal Processing*, 93(9):2696–2708. DOI: 10.1016/j.sigpro.2013.03.005. 79, 80

Wang, S., Zhang, D., Liang, Y., and Pan, Q. (2012). Semi-coupled dictionary learning with applications to image super-resolution and photo-sketch synthesis. In *Computer Vision and Pattern Recognition (CVPR), 2012 IEEE Conference on*, pages 2216–2223. IEEE. DOI: 10.1109/CVPR.2012.6247930. 52, 88

Wang, X., Wang, B., Bai, X., Liu, W., and Tu, Z. (2013c). Max-margin multiple-instance dictionary learning. In *Proceedings of The 30th International Conference on Machine Learning*, pages 846–854. 94

Wang, Z., Bovik, A. C., Sheikh, H. R., and Simoncelli, E. P. (2004). Image quality assessment: from error visibility to structural similarity. *Image Processing, IEEE Transactions on*, 13(4):600–612. DOI: 10.1109/TIP.2003.819861. 80

Wen, Z., Yin, W., Zhang, H., and Goldfarb, D. (2011). On the convergence of an active-set method for ℓ_1 minimization. *Optimization Methods and Software*. DOI: 10.1080/10556788.2011.591398. 48

Wright, J., Ganesh, A., Rao, S., Peng, Y., and Ma, Y. (2009a). Robust principal component analysis: Exact recovery of corrupted low-rank matrices by convex optimization. In *Proc. of Neural Information Processing Systems*, volume 3. 14, 68, 69, 91, 92, 93, 95

Wright, J., Ganesh, A., Rao, S., Peng, Y., and Ma, Y. (2009b). Robust principal component analysis: Exact recovery of corrupted low-rank matrices via convex optimization. In Bengio, Y., Schuurmans, D., Lafferty, J., Williams, C. K. I., and Culotta, A., editors, *Advances in Neural Information Processing Systems 22*, pages 2080–2088. 46, 47

Wright, J., Yang, A. Y., Ganesh, A., Sastry, S. S., and Ma, Y. (2009c). Robust face recognition via sparse representation. *Pattern Analysis and Machine Intelligence, IEEE Transactions on*, 31(2):210–227. DOI: 10.1109/TPAMI.2008.79. 9, 38, 90, 94, 103, 104, 107

Wright, S. J., Nowak, R. D., and Figueiredo, M. A. (2009d). Sparse reconstruction by separable approximation. *Signal Processing, IEEE Transactions on*, 57(7):2479–2493. DOI: 10.1109/TSP.2009.2016892. 11

Wu, Y., Ling, H., Blasch, E., Bai, L., and Chen, G. (2011). Visual tracking based on log-euclidean riemannian sparse representation. In *Advances in Visual Computing*, pages 738–747. Springer. DOI: 10.1007/978-3-642-24028-7_68. 99, 101

Xiang, Z. J., Xu, H., and Ramadge, P. J. (2011). Learning sparse representations of high dimensional data on large scale dictionaries. In *NIPS*, volume 24, pages 900–908. 56

Xie, B., Song, M., and Tao, D. (2010). Large-scale dictionary learning for local coordinate coding. In *BMVC*, pages 1–9. DOI: 10.5244/C.24.36. 62

Xie, Z. and Feng, J. (2009). Kfce: A dictionary generation algorithm for sparse representation. *Signal Processing*, 89(10):2072–2077. DOI: 10.1016/j.sigpro.2009.04.001. 27

Xing, J., Gao, J., Li, B., Hu, W., and Yan, S. (2013). Robust object tracking with online multi-lifespan dictionary learning. In *Computer Vision (ICCV), 2013 IEEE International Conference on*, pages 665–672. IEEE. DOI: 10.1109/ICCV.2013.88. 62, 100

Yaghoobi, M., Blumensath, T., and Davies, M. E. (2009a). Dictionary learning for sparse approximations with the majorization method. *Signal Processing, IEEE Transactions on*, 57(6):2178–2191. DOI: 10.1109/TSP.2009.2016257. 29, 77

Yaghoobi, M., Blumensath, T., and Davies, M. E. (2009b). Parsimonious dictionary learning. In *Acoustics, Speech and Signal Processing, 2009. ICASSP 2009. IEEE International Conference on*, pages 2869–2872. IEEE. DOI: 10.1109/ICASSP.2009.4960222. 29, 77

Yaghoobi, M., Daudet, L., and Davies, M. E. (2009c). Parametric dictionary design for sparse coding. *Signal Processing, IEEE Transactions on*, 57(12):4800–4810. DOI: 10.1109/TSP.2009.2026610. 31

Yaghoobi, M., Nam, S., Gribonval, R., and Davies, M. E. (2012). Noise aware analysis operator learning for approximately cosparse signals. In *Acoustics, Speech and Signal Processing (ICASSP), 2012 IEEE International Conference on*, pages 5409–5412. IEEE. DOI: 10.1109/ICASSP.2012.6289144. 78, 80

Yan, J., Zhu, M., Liu, H., and Liu, Y. (2010). Visual saliency detection via sparsity pursuit. *Signal Processing Letters, IEEE*, 17(8):739–742. DOI: 10.1109/LSP.2010.2053200. 9, 95, 96

Yang, J., Wang, Z., Lin, Z., Cohen, S., and Huang, T. (2012). Coupled dictionary training for image super-resolution. *Image Processing, IEEE Transactions on*, 21(8):3467–3478. DOI: 10.1109/TIP.2012.2192127. 9

Yang, J., Wright, J., Huang, T., and Ma, Y. (2008). Image super-resolution as sparse representation of raw image patches. In *Computer Vision and Pattern Recognition, 2008. CVPR 2008. IEEE Conference on*, pages 1–8. IEEE. DOI: 10.1109/CVPR.2008.4587647. 51, 52, 53, 68, 87, 88, 89

Yang, J., Wright, J., Huang, T. S., and Ma, Y. (2010a). Image super-resolution via sparse representation. *Image Processing, IEEE Transactions on*, 19(11):2861–2873. DOI: 10.1109/TIP.2010.2050625. 51, 52, 53, 69, 87, 88, 89

Yang, J. and Yang, M.-H. (2012). Top-down visual saliency via joint crf and dictionary learning. In *Computer Vision and Pattern Recognition (CVPR), 2012 IEEE Conference on*, pages 2296–2303. IEEE. DOI: 10.1109/CVPR.2012.6247940. 67, 97

Yang, J., Yu, K., Gong, Y., and Huang, T. (2009). Linear spatial pyramid matching using sparse coding for image classification. In *Computer Vision and Pattern Recognition, 2009. CVPR 2009. IEEE Conference on*, pages 1794–1801. IEEE. DOI: 10.1109/CVPRW.2009.5206757. 9

Yang, J., Yu, K., and Huang, T. (2010b). Efficient highly over-complete sparse coding using a mixture model. In *Computer Vision–ECCV 2010*, pages 113–126. Springer. DOI: 10.1007/978-3-642-15555-0_9. 66

Yang, J. and Zhang, Y. (2011). Alternating direction algorithms for \ell_1-problems in compressive sensing. *SIAM Journal on Scientific Computing*, 33(1):250–278. DOI: 10.1137/090777761. 11

Yang, M., Dai, D., Shen, L., and Gool, L. (2013a). Latent dictionary learning for sparse representation based classification. In *Proceedings of the IEEE Conference on Computer Vision and Pattern Recognition*, pages 4138–4145. DOI: 10.1109/CVPR.2014.527. 37, 42

Yang, M., Zhang, D., and Feng, X. (2011a). Fisher discrimination dictionary learning for sparse representation. In *Computer Vision (ICCV), 2011 IEEE International Conference on*, pages 543–550. IEEE. DOI: 10.1109/ICCV.2011.6126286. 42

Yang, M. and Zhang, L. (2010). Gabor feature based sparse representation for face recognition with gabor occlusion dictionary. In *Computer Vision–ECCV 2010*, pages 448–461. Springer. DOI: 10.1007/978-3-642-15567-3_33. 30

Yang, S., Liu, Z., Wang, M., Sun, F., and Jiao, L. (2011b). Multitask dictionary learning and sparse representation based single-image super-resolution reconstruction. *Neurocomputing*, 74(17):3193–3203. DOI: 10.1016/j.neucom.2011.04.014. 52, 53, 88, 89

Yang, S., Min, W., Zhao, L., and Wang, Z. (2013b). Image noise reduction via geometric multi-scale ridgelet support vector transform and dictionary learning. *Image Processing, IEEE Transactions on*, 22(11):4161–4169. DOI: 10.1109/TIP.2013.2271114. 58, 80

Yang, Y. (2005). Information theory, inference, and learning algorithms. *Journal of the American Statistical Association*, 100(472):1461–1462. DOI: 10.1198/jasa.2005.s54. 6

Yao, B., Jiang, X., Khosla, A., Lin, A. L., Guibas, L., and Fei-Fei, L. (2011). Human action recognition by learning bases of action attributes and parts. In *Computer Vision (ICCV), 2011 IEEE International Conference on*, pages 1331–1338. IEEE. DOI: 10.1109/ICCV.2011.6126386. 30

Yin, W., Osher, S., Goldfarb, D., and Darbon, J. (2008). Bregman iterative algorithms for \ell_1-minimization with applications to compressed sensing. *SIAM Journal on Imaging Sciences*, 1(1):143–168. DOI: 10.1137/070703983. 11

Yu, K., Lin, Y., and Lafferty, J. (2011). Learning image representations from the pixel level via hierarchical sparse coding. In *Computer Vision and Pattern Recognition (CVPR), 2011 IEEE Conference on*, pages 1713–1720. IEEE. DOI: 10.1109/CVPR.2011.5995732. 55

Yuan, M. and Lin, Y. (2006). Model selection and estimation in regression with grouped variables. *Journal of the Royal Statistical Society: Series B (Statistical Methodology)*, 68(1):49–67. DOI: 10.1111/j.1467-9868.2005.00532.x. 12

Zayyani, H. and Babaie-Zadeh, M. (2009). Thresholded smoothed-l0 (sl0) dictionary learning for sparse representations. In *Acoustics, Speech and Signal Processing, 2009. ICASSP 2009. IEEE International Conference on*, pages 1825–1828. IEEE. DOI: 10.1109/ICASSP.2009.4959961. 28

Zeyde, R., Elad, M., and Protter, M. (2012). On single image scale-up using sparse-representations. In *Curves and Surfaces*, pages 711–730. Springer. DOI: 10.1007/978-3-642-27413-8_47. 88, 89

Zhang, G., Jiang, Z., and Davis, L. S. (2013a). Online semi-supervised discriminative dictionary learning for sparse representation. In *Computer Vision–ACCV 2012*, pages 259–273. Springer. DOI: 10.1007/978-3-642-37331-2_20. 62

Zhang, H., Yang, J., Zhang, Y., Nasrabadi, N. M., and Huang, T. S. (2011). Close the loop: Joint blind image restoration and recognition with sparse representation prior. In *Computer Vision (ICCV), 2011 IEEE International Conference on*, pages 770–777. IEEE. DOI: 10.1109/ICCV.2011.6126315. 85

Zhang, H., Zhang, Y., and Huang, T. S. (2013b). Simultaneous discriminative projection and dictionary learning for sparse representation based classification. *Pattern Recognition*, 46(1):346–354. DOI: 10.1016/j.patcog.2012.07.010. 40, 94

Zhang, K., Gao, X., Tao, D., and Li, X. (2012a). Multi-scale dictionary for single image super-resolution. In *Computer Vision and Pattern Recognition (CVPR), 2012 IEEE Conference on*, pages 1114–1121. IEEE. DOI: 10.1109/CVPR.2012.6247791. 87, 89

Zhang, K., Zhang, L., and Yang, M.-H. (2012b). Real-time compressive tracking. In *Computer Vision–ECCV 2012*, pages 864–877. Springer. DOI: 10.1007/978-3-642-33712-3_62. 101

Zhang, Q. and Li, B. (2010a). Discriminative k-svd for dictionary learning in face recognition. In *Computer Vision and Pattern Recognition (CVPR), 2010 IEEE Conference on*, pages 2691–2698. IEEE. DOI: 10.1109/CVPR.2010.5539989. 1, 34, 37, 38, 94, 103, 107

Zhang, Q. and Li, B. (2010b). Joint sparsity model with matrix completion for an ensemble of face images. In *Image Processing (ICIP), 2010 17th IEEE International Conference on*, pages 1665–1668. IEEE. DOI: 10.1109/ICIP.2010.5650188. 46

Zhang, Q. and Li, B. (2012). Mining discriminative components with low-rank and sparsity constraints for face recognition. In *Proceedings of the 18th ACM SIGKDD international conference on Knowledge discovery and data mining*, pages 1469–1477. ACM. DOI: 10.1145/2339530.2339760. 14, 46, 49

Zhang, Q., Zhou, J., Wang, Y., Ye, J., and Li, B. (2014). Image cosegmentation via multi-task learning. In *British Machine Vision Conference*. 90, 91

Zhang, S., Yao, H., Sun, X., and Liu, S. (2010). Robust object tracking based on sparse representation. In *Visual Communications and Image Processing 2010*, pages 77441N–77441N. International Society for Optics and Photonics. DOI: 10.1117/12.863437. 102

Zhang, S., Yao, H., Zhou, H., Sun, X., and Liu, S. (2013c). Robust visual tracking based on online learning sparse representation. *Neurocomputing*, 100:31–40. DOI: 10.1016/j.neucom.2011.11.031. 101

Zhang, T., Ghanem, B., Liu, S., and Ahuja, N. (2012c). Robust visual tracking via multi-task sparse learning. In *Computer Vision and Pattern Recognition (CVPR), 2012 IEEE Conference on*, pages 2042–2049. IEEE. DOI: 10.1109/CVPR.2012.6247908. 62, 101

Zhang, W., Wang, X., Zhao, D., and Tang, X. (2012d). Graph degree linkage: Agglomerative clustering on a directed graph. In *Computer Vision–ECCV 2012*, pages 428–441. Springer. DOI: 10.1007/978-3-642-33718-5_31. 6

Zhang, W., Zhao, D., and Wang, X. (2013d). Agglomerative clustering via maximum incremental path integral. *Pattern Recognition*, 46(11):3056–3065. DOI: 10.1016/j.patcog.2013.04.013. 6

Zhao, B., Fei-Fei, L., and Xing, E. P. (2011). Online detection of unusual events in videos via dynamic sparse coding. In *Computer Vision and Pattern Recognition (CVPR), 2011 IEEE Conference on*, pages 3313–3320. IEEE. DOI: 10.1109/CVPR.2011.5995524. 97

Zhao, M., Li, S., and Kwok, J. (2010). Text detection in images using sparse representation with discriminative dictionaries. *Image and Vision Computing*, 28(12):1590–1599. DOI: 10.1016/j.imavis.2010.04.002. 90

Zhao, P., Rocha, G., and Yu, B. (2009). The composite absolute penalties family for grouped and hierarchical variable selection. *The Annals of Statistics*, pages 3468–3497. DOI: 10.1214/07-AOS584. 12, 54

Zhao, Z., Ahn, G.-J., Seo, J.-J., and Hu, H. (2013). On the security of picture gesture authentication. In *USENIX Security*, pages 383–398. 95

Zheng, J. and Jiang, Z. (2013). Tag taxonomy aware dictionary learning for region tagging. In *Computer Vision and Pattern Recognition (CVPR), 2013 IEEE Conference on*, pages 369–376. IEEE. DOI: 10.1109/CVPR.2013.54. 57

Zheng, J., Jiang, Z., Phillips, P. J., and Chellappa, R. (2012). Cross-view action recognition via a transferable dictionary pair. In *BMVC*, volume 1, page 7. DOI: 10.5244/C.26.125. 53

Zheng, M., Bu, J., Chen, C., Wang, C., Zhang, L., Qiu, G., and Cai, D. (2011). Graph regularized sparse coding for image representation. *Image Processing, IEEE Transactions on*, 20(5):1327–1336. DOI: 10.1109/TIP.2010.2090535. 31

Zhou, M., Chen, H., Paisley, J., Ren, L., Li, L., Xing, Z., Dunson, D., Sapiro, G., and Carin, L. (2012a). Nonparametric bayesian dictionary learning for analysis of noisy and incomplete images. *Image Processing, IEEE Transactions on*, 21(1):130–144. DOI: 10.1109/TIP.2011.2160072. 71, 80

Zhou, M., Chen, H., Ren, L., Sapiro, G., Carin, L., and Paisley, J. W. (2009). Non-parametric bayesian dictionary learning for sparse image representations. In *Advances in Neural Information Processing Systems*, pages 2295–2303. 17, 63, 66, 70, 71, 80, 82

Zhou, M., Yang, H., Sapiro, G., Dunson, D. B., and Carin, L. (2011). Dependent hierarchical beta process for image interpolation and denoising. In *International Conference on Artificial Intelligence and Statistics*, pages 883–891. 71, 80

Zhou, N., Shen, Y., Peng, J., and Fan, J. (2012b). Learning inter-related visual dictionary for object recognition. In *Computer Vision and Pattern Recognition (CVPR), 2012 IEEE Conference on*, pages 3490–3497. IEEE. 50

Zhou, S. K., Chellappa, R., and Moghaddam, B. (2004). Visual tracking and recognition using appearance-adaptive models in particle filters. *Image Processing, IEEE Transactions on*, 13(11):1491–1506. DOI: 10.1109/TIP.2004.836152. 100

Zhou, Y. and Barner, K. (2013). Locality constrained dictionary learning for nonlinear dimensionality reduction. *Signal Processing Letters, IEEE*, 20(4):335–338. DOI: 10.1109/LSP.2013.2246513. 31

Zontak, M. and Irani, M. (2011). Internal statistics of a single natural image. In *Computer Vision and Pattern Recognition (CVPR), 2011 IEEE Conference on*, pages 977–984. IEEE. DOI: 10.1109/CVPR.2011.5995401. 73

Zou, H. and Hastie, T. (2005). Regularization and variable selection via the elastic net. *Journal of the Royal Statistical Society: Series B (Statistical Methodology)*, 67(2):301–320. DOI: 10.1111/j.1467-9868.2005.00503.x. 11

Zou, H., Hastie, T., and Tibshirani, R. (2006). Sparse principal component analysis. *Journal of computational and graphical statistics*, 15(2):265–286. DOI: 10.1198/106186006X113430. 11

Authors' Biographies

QIANG ZHANG

Qiang Zhang received his B.S. degree in electronic information and technology from Beijing Normal University, Beijing, China in 2009 and his Ph.D. degree in Computer Science from Arizona State University, Tempe, Arizona in 2014. Since 2014, he has been with Samsung, Pasadena, CA as a staff research scientist in computer vision and machine learning. His research interests include image/video processing, computer vision and machine vision, specialized in sparse learning, face recognition, and motion analysis.

BAOXIN LI

Baoxin Li received his Ph.D. in electrical engineering from the University of Maryland, College Park, in 2000. He is currently a professor of computer science and engineering and a graduate faculty in computer science, electrical engineering and computer engineering programs at Arizona State University, Tempe. From 2000 to 2004, he was a Senior Researcher with SHARP Laboratories of America, Camas, Washington, where he was a technical lead in developing SHAR'sP HiMPACT Sports technologies. From 2003–2004, he was also an Adjunct Professor with the Portland State University, Oregon. He holds sixteen issued U.S. patents and his current research interests include computer vision and pattern recognition, multimedia, social computing, machine learning, and assistive technologies. He won twice the SHARP Laboratories' President Award, in 2001 and 2004 respectively. He also won the SHARP Laboratories' Inventor of the Year Award in 2002. He was a recipient of the National Science Foundation's CAREER Award.

Printed in the United States
by Baker & Taylor Publisher Services